DPH MATHEMATICS SERIES

TEXT BOOK OF LINEAR PROGRAMMING

I

By

A.K. Sharma

DISCOVERY PUBLISHING HOUSE
NEW DELHI-110002

Reprinted - 2019

First Published - 2005

ISBN: 978-81-8536-002-3

Text book of Linear Programming - I

Published by:

DISCOVERY PUBLISHING HOUSE PVT. LTD.
4383/4B, Ansari Road, Darya Ganj
New Delhi-110 002 (India)
Phone: +91-11-23279245, 43596064-65
Fax: +91-11-23253475
E-mail: discoverypublishinghouse@gmail.com
sales@discoverypublishinggroup.com
web: www.discoverypublishinggroup.com

Printed at:
Infinity Imaging Systems
Delhi

Preface

Linear Programming has progressed a great deal during last two decades. It is becoming increasingly sophisticated with the availability of computer facilities and infusion of new concepts. The text of this book has been presented in easy and simple language. Throughout the text, the two streams theory and technique run side by side. Each technique is preceded by the relevant theory followed by suitable examples. A large number of important problems mostly drawn from university examination papers has been included.

Every efforts has been made to keep the book free from printing errors even so some might have escaped our notice. We shall be highly obliged to the reader who may point out any such error to us. Suggestions for improvement of the book will be gratefully received.

A.K. Sharma

Contents

TIME MINIMIZATION PROBLEM

In the Time Minimizing Transportation Problem, the objective is to minimize the time of transportation rather than the cost of transportation. Such problems are often encountered in military, hospital, management, five service etc.

The minimization transportation problems are similar to the problems discussed for, except that the unit time t_{ij} required to transport the items from the origin to the destination.

The systematic procedure for the solution of such problems is as follows:

Step 1. Determine an initial basic feasible solution by using any methods.

Step 2. Determine ^{T}j for this basic feasible solution and cross out all the non basic calls for which $t_{ij} \geq {}^{T}j$

Step 3. Construct a loop for the basic cells corresponding to T_j in such a way that when the values cut the corner cells are shifted around, the value this cell tends towards (not necessarily zero) zero and no variable becomes zero. It no such closed path can be formed, the solution under test is optimal, otherwise move to next step.

Step 4. Go to step 2 and repeat the procedure until an optimum basic feasible solution is attained.

TRANSSHIP MINT PROBLEM

A transportation problem in which available commodity may not be sent directly from which available commodity may not be sent directly from sources to destination is termed as a transshipment problem. Such

[2] 7	3	4	2
[7] 7	[1] 1	[6] 3	5t ∈ = 5
3	4	[5] 6	
4	1	5t ∈ = ∈	

Notice that even after the inclusion of cell (2, 3) in the basis, the basic cells do not form a loop. The net evaluation can now be computed and solution is tested for optimality.

STARTING TABLE

Since all the net evaluations for the non basic variables are not non positive, the initial solution is not optimum. The non basic call (1, 3) must enter the basis: The exist criterion removes basis cell (2, 3) from the basis [Max. $\theta = \in$]

			u_i
2 $-\theta$ 7	(3) 3	θ (4) 4	5
2 $+\theta$ 2	1 1	$\in$ $-\theta$ 3	0
(2) 3	(0) 4	5 6	3
v_j 2	1	2	

Starting table

First Iteration

Induced the cell (1, 3) into the basis and drop the cell (2, 3) from it Determine the current net evaluations. Since all of them are not non positive, the current solution can be improved.

			u_i
2 $-\theta$ 7	(3) 3	$\in$ $+\theta$ 4	0
2 2	1 1	(–4) 3	–5
θ (6) 5	(4) 4	5 $-\theta$ 6	2
v_j 7	6	4	

First iterated table

Second Iteration

Introduced the cell (3, 1) and drop the cell (1, 1) from the basis since some of the current net evolution are still positive, the current solution can for there be improved.

			u_i
(–6) 7	(–3) (3)	2 4	–1
2 –θ 2	1 1	θ (2) (3)	0
2 +θ 3	(–2) (4)	3 –θ 6	2
v_j 2	1	5	

Second Iteration table:

Third Iteration

Introduced the cell (2, 3) and drop the cell (2, 1) from the basis. Since all the current net evaluation are non positive the current solution is an optimum one.

			u_i
(–6) 7	(–1) (3)	2 4	4
(–2) 2	1 1	2 3	3
4 5	(0) (4)	1 6	6
v_j –3	–2	0	

Optimum table.

The Transportation cost according to above route is given by

Z = 2 × 4 + 1 × 1 + 2 × 3 + 4 × 3 + 1 × 6 =33

UNBALANCED TRANSPORTATION PROBLEM

Some time there is more demand than the availability and *vice- versa* such type of transportation problem are called unbalanced transportation problems.

If in a transportation problem

$$\sum_{i=1}^{m} ai \neq \sum_{i=1}^{n} bj$$

Then the problem is called unbalances transportation problem. There are two type of unbalance problem.

Case I: (Excess Availability) : When never $\Sigma a_i - \Sigma b_j$ we introduced a dummy destination in a transportation table the cost of transporting to this destination are all set equal to zero. The requirement at this dummy destination is assumed to be equal to $\Sigma a_i - \Sigma b_j$

Example 1:

Solve the transportation table.

	D_1	D_2	D_3	*available*
	20	*30*	*10*	*800*
	5	*15*	*25*	*500*
Required	*300*	*300*	*400*	

Solution:

Here the total available exceed the total demand. We consider a fictitious destination requiring $\sum_{i=1}^{m} a_i - \sum_{j=1}^{n} b_j = 1300 - 1000 = 300$ units of quantity. The new transportation problem is

	D_1	D_2	D_3	fictitious	available
O_1	30	20	10	0	800
O_2	5	15	25	0	500
Required	300	300	400	300	1300

Solve this by usual method; we get are optimal solution given by the following table

	D_1	D_2	D_3	fictitious	available
O_1		100	400	300	800
O_2	300	200			500
Required	300	300	400	300	

The table indicate that the 300 units will be less at O_1 not dispatched.

Case II: (Shortage in Availability) : Whenever $\Sigma a_i \leq \Sigma b_j$, introduced a dummy source in the transportation table. The cost of transporting from this source to any destination any all set equal to zero the availability at this dummy source is assumed to be equal to

$$\Sigma b_j - \Sigma a_i$$

to or (subtracted from) a non zero quantity, we need to write Σ only when it appears alone.

Remark:

A degenerate B.F.S. can also be converted into a non degenerate B.F.S. by introducing a small positive quantity say I in one or more empty (independent) cells and correspondingly increasing the capacity and demand by the same quantity say in one or more empty (independent) cells and correspondingly increasing the capacity and demand by the same quantity 1.

Example 2:

Obtain an optimum basic feasible solution to the following degenerate T.P.

	To				
	7	*3*	*4*	*2*	
from	*2*	*1*	*3*	*3*	*available*
	3	*4*	*6*	*5*	
Demand.	*4*	*1*	*5*	*10*	

Solution:

The initial solution by north west carnivore is

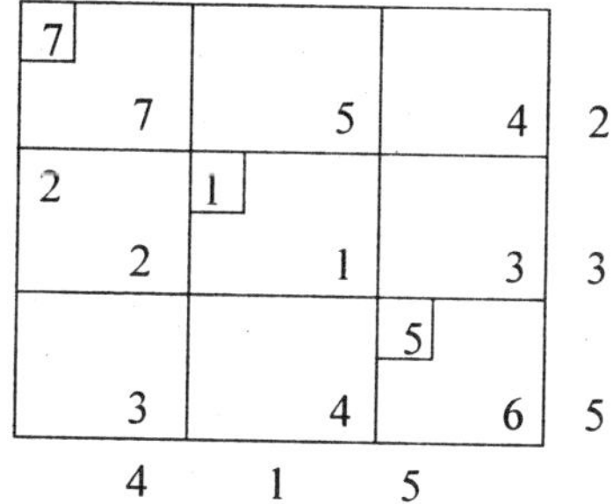

In order to complete the basic and there by remove degeneracy we rigger only one more positive basic variable we select the variable x_{23} and allocate a negligibly small positive quantity in the call (2, 3) as shown in the table below.

Degeneracy in Transportation Problem

A feasible solution to a transportation problem is degenerate if less than m + n – 1 of the x_{ij} are positive. Degeneracy may be encountered in the process of determining the initial basic feasible solution or at some subsequent iteration. The reason for degeneracy is that the sum of certain number of a_i is exactly equal to the sum of some b_j is . Due to this property, while finding a F.S. the allocation in some cell, satisfies the column as well as row requirements simultaneously.

In such case we chose one or more as many individual allocations are required to have exactly m + n – 1 allocations) empty cell of the matrix (generally lowest cost cells is suits) in a way that if these calls are allocated, they form an independent set of allocations. In these chosen cell or cells, we may assign one or very small positive quantity '$\in$' (epsilon), governed by the following rules:

(i) $O < \in < x_{ij}$

(ii) $O + \in = \in$

(iii) $x_{ij} \pm \in = x_{ij}$

Further, as long as it is needed, this $\in$ is to be considered as a real allocation through all the subsequent iteration. Above rule show that even after introducing $\in$, the original problem is not changed. It is merely a technique to apply the optimality test. As $\in$ has no physical significant ultimately it is to be omitted.

In numerical problem due to the property

(3) Of $\in$, we need not to write it whenever it is added.

THE TRANSPORTATION ALGORITHM

The Algorithm

Various steps are as follows :

Step 1. Construct a transportation table, entering the origin capacities a_i, the destination requirements b_j and the cost c_{ij}.

Step 2. Find the initial basic feasible solution. Enter the solution in the upper left corners of the basic cells.

Step 3. Find the set of numbers u_i and ui for each row and column s.t. for all the occupied cells $C_{ij} = u_i + v_j \; \forall_{i,j}$ for which (i, j) is in the basis starting initially with some $u_i = 0$ and entering successively the value of u_i and v_j on the transportation table.

Step 4. Computer the net evaluation $Z_{ij} - c_{ij} = u_i + v_j - c_{ij}$ for all the non basic cells and enter them in the upper right corners of the corresponding cells.

Step 5. Examine the sign of each $Z_{ij} - c_{ij}$. It all $Z_{ij} - c_{ij} \leq 0$, then the current boric feasible solution is an optimum one. It at least one $Z_{ij} - c_{ij} > 0$, select the variable x_{rs} having the largest positive net evaluation to enter the basis.

Step 6. Let the variable x_{rs} enter the basis. Allocate an unknown quantity say θ, to the cell (r, s). Identity a loop that an starts and end at the cell (r, s) and connect some of the basic cells. Add and subtract interchangeably, θ to and from the transition cells of the loop in such a way that the rim requirements remain satisfied.

Step 7. Assign a maximum value θ in such a way that the value of one basic variable be comes zero and the other bane variables remain non-negative. The basic cell whose allocation has been reduced to zero, leaves the basis.

Step 8. Return to step 3 and repeat the process until an optimum basic feasible solution has been obtained.

Note:

This procedure is known as u – v method or MODI (Modified Distribution) method.

The various steps of the algorithm are summarized in flow chart.

Transportation Algorithm for minimization T.P.

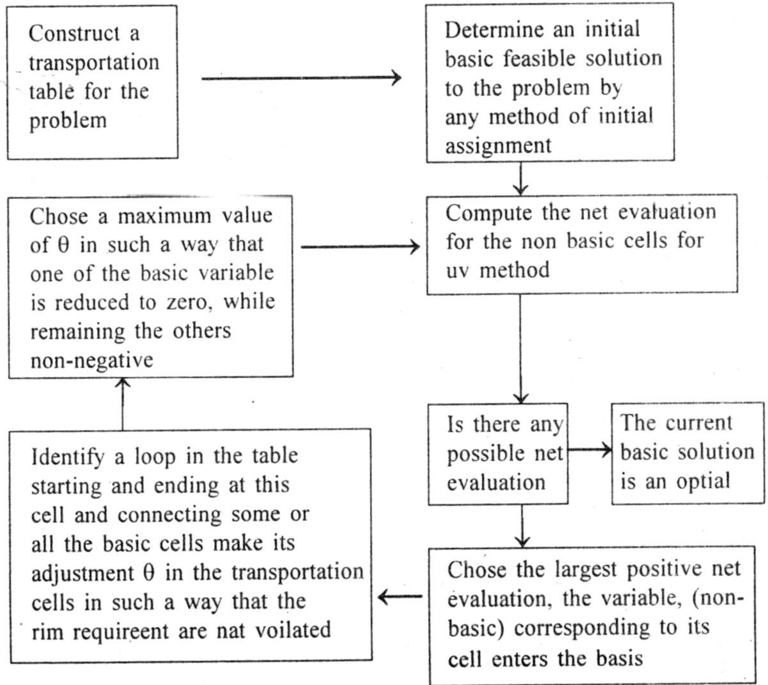

The Column Minima Method

Step 1. Determine the smallest cost in the first column of the transportation table. Let it be c_{ij}. Allocate x_{ij} = min (a_1, b_1,) in the call (i, 1)

Step 2. If $x_{ij} = b_1$, cross off the first column of the transportation table and move towards right to the second column.

If $yi_1 = ai_1$ cross off the ith row of the transportation table and reconsider the first column with the remaining demand.

If $xj_1 = b_1 = ai$, cross off the ith row and make the second allocation $yk_1 = 0$ in the cell (k, 1) with ck_1 being the new minimum cost in the first column. Cross of the column and move towards right to the second column.

Step 3. Repeat step 1 and 2 for the resulting reduced transportation table until all the rim requirements are satisfied.

The Row Minima Method

Step 1. The smallest cost in the first row of the transportation table determined. Let it be C_{ij}. Allocate the maximum decibel amount x_{ij} = min (a, bj) is the cell (i, j) so that either the capacity of origin O_1 is exhausted, or the requirement at destination D_i is satisfied or both.

Step 2. If $x_{1j} = a_1$ so that the availability at origin O_1 is completely exhausted, cross off the first row of the table and move down to the second now.

If $x_{ij} = b_j$ so that the requirement at destination as satisfied, cross off the jth column and reconsider the first row with the remaining availability of origin O_1.

If $x_{ij} = a_i = b_j$ the origin capacity O_1 is completely exhausted as well as the requirement at destination D_j is completely satisfied an arbitrary tie breaking choice is made. Cross off the jth column and made the second allocation $x_{ik} = 0$ in the cell (1, k) with c_1k being the new minimum cost in the first row. Cross off the first row and move down to the second row.

Step 3. Repeat step 1 and 2 for the resulting reduced transportation table until all the rim requirements are satisfied.

TRIANGULAR BASIS IN T.P.

The reader may recall that a system of n linear equations Ax = b is called a triangular system it the matrix A is triangular i.e., it is the form.

$$A = \begin{pmatrix} a_{11} & a_{12} & a_{13} & a_{1n} \\ 0 & a_{22} & a_{23} & a_{2n} \\ 0 & 0 & a_{33} & a_{3n} \\ 0 & 0 & 0 & a_{nn} \end{pmatrix} a_{ij} \neq 0 \text{ for each i}$$

The system has a property that there is at least one equation having only one variable. This variable can be eliminated from the remaining equations resulting in a reduced system of (n – 1) equations which again happens to be triangular. Thus a triangular system can be easily solved by back substitution.

Definition: (Triangular Basis) : A basis for the system Ax = b is said to be a triangular basis if a Triangular system is obtained when all the non basic variables are set zero in the system.

LOOPS IN TRANSPORTATION TABLE AND THEIR PROPERTIES

Definition (Loop): In a Transportation table, an order set of four or more cells is said to form a loop if : (i) any two adjacent cells in the ordered set lie either in the same row or in the same column and (ii) any three or more adjacent cells in the ordered set do not lie in the same row or the same column the first call of the set is considered to follow the last in the set.

If we join the cells of a loop by horizontal and vertical segments we get a closed path satisfying : (i) and (ii) condition. Let us denote by (i, j) the (i, j)th cell of a transportation table. Then, it can be seen from the graphical illustration in the fig. given below: tha' the set

L = {(2, 1), (4, 1), (4, 4), (1, 4), (1, 2), (2, 2)}

forms at loop while the set

L' = {(3, 2), (3, 5), (2, 5), (2, 4), (2, 3), (1, 3), (1, 2)}

does not form a loop.

Definition: 2. (Set Containing a Loop) : A set x of cells of a transportation table is said to CONTAIN A LOOP if a call of x or of a subset of x can be ordered (sequenced) so as to form a loop.

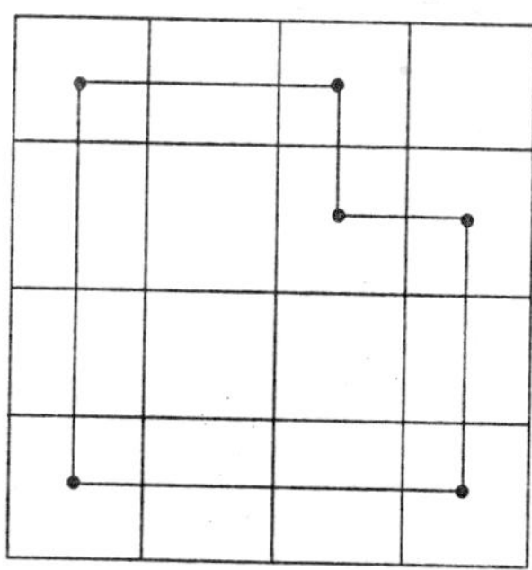

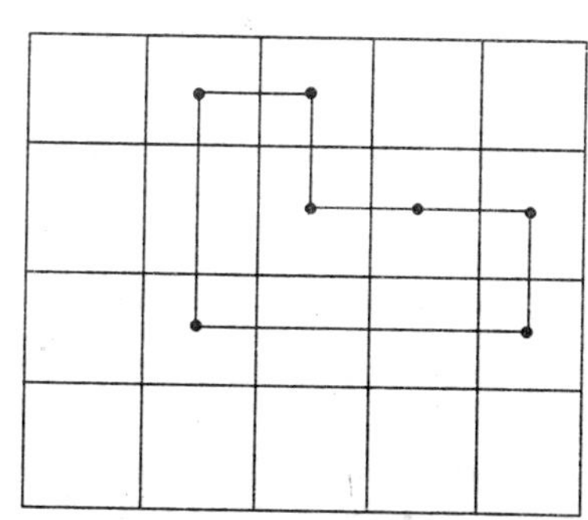

Where $x = [x_{12} \ldots, x_{1n}, x_{21} \ldots, x_{2n} \ldots, x_{2n} \ldots, x_{m1} \ldots x_{mn}]$ C is the cost vector $b = \{a_1 \ldots a_m, b_1 \ldots b_n\}$ and A is an $(m + n) \times mn$ real matrix containing the coefficient of constraints.

The reader should note that the elements of A are either 0 or +1 thus the general L.P.P can be reduced to be a T.P. if

(i) The a_{ij}'s are restricted to the value of 0 and + 1, and

(ii) The units among the constraints are homogeneous for a T.P. involving 2 origins and 3 destination $(m = 2, n = 3)$ the matrix A is given by

$$A = \begin{bmatrix} 1 & 1 & 1 & 0 & 0 & 0 \\ 0 & 0 & 0 & 1 & 1 & 1 \\ \hline 1 & 0 & 0 & 1 & 0 & 0 \\ 0 & 1 & 0 & 0 & 1 & 0 \\ 0 & 0 & 1 & 0 & 0 & 1 \end{bmatrix} = \begin{bmatrix} e^1 23 & e^2 23 \\ I_3 & I_3 \end{bmatrix}$$

and therefore, in general for an m origin, n destination T. P we may write

$$A = \begin{bmatrix} e^1 mn & e^2 mn \ldots & e^m mn \\ In & In \ldots & In \end{bmatrix}$$

where e'mn is an $m \times n$ matrix having a row of write elements as its jth row and o's every where else and in is the $n \times n$ Identity matrix.

If aij denotes the column vector at A associated with any variable x_{ij} then it easily verified that $a_{ij} = e_j + e_{mj} - e_{ij}$ where $e_j, e_{m+j} \in R^{m+n}$ are unit vectors.

Subtracting the two, we have

$$\sum_{i=1}^{m}\sum_{j=1}^{n} xij - \sum_{j=1}^{n-1}\sum_{i=1}^{m} xij = \sum_{j=1}^{m} ai - \sum_{j=1}^{n-1} bj$$

or
$$\sum_{i=1}^{m}\left(\sum_{j=1}^{n} xij - \sum_{j=1}^{n-1} xij\right) = \sum_{i=1}^{n} bi - \sum_{j=1}^{n-1} bi \quad (\in a_i = \in b) \ (\because \sum_j ai = \sum_j b)$$

or
$$\sum_{i=1}^{m} xin = bn$$

which happens to be the last destination constraint of the problem.

This indicates that it the first m + n – 1 constraints are satisfied then $\Sigma\, a_i = \Sigma b_j$ ensures that the (m + n)th constraint will be automatically satisfied.

This, out m + n equation, we have only (m + n – 1) linearly independent equations. Therefore, a basic feasible solution will consist of at most (m + n – 1) positive variables the rest being zero. Further a feasible solution involving exactly (m + n – 1) positive variables is known as non degenerate basic feasible solution otherwise it is said to be degenerate basic feasible.

MATRIX FORM OF T.P.

Consider the T.P. discussed in the last section. The set of constraints $\sum_{j=1}^{n} xij = ai$ and $\sum_{j=1}^{m} xij = b_j$ represent m + n equation in mn non negative variable x_{ij} each variable x_{ij} appears in exactly tow constraints one associated with the ith origin O_i and other with the jth destination D_j. In the above ordering of constraints – the origin equation first, then the destination equation – the T. P. can be restated in the matrix from as.

Minimize $z = c^T \times c'\ x \in R^{mn}$ subject to the constraints

$Ax = b,\ x \geq 0\ b \in R^{m+n}$

Corollary (Existence of an Optimal Solution)

There always exists an optimal solution to a balanced transportation problem.

Proof:

Let $\sum_j a_i = \sum_j b$ so that a feasible solution x_{ij} exists. Therefore follows from the constraints of the problem that each x_{ij} is bounded

viz $0 \leq x_{ij} \leq \min(a, b)$

Thus the feasible region of the problem is closed, bounded and non empty and hence there exist an optimal solution.

Note:

In future we shall resume that the above condition holds for the T.P. we are discussing without mentioning it.

BASIC FEASIBLE SOLUTION OF A TRANSPORTATION PROBLEM

Since a T.P. is a spacial case of L.P.P. a basic feasible solution a transportation problem has the same definition as in for an L.P.P. However, we observe that in the case of a T.P. there are only m + n – 1 basic variable out of mn unknown, due to reducing in the constraints of the T.P.

For consider the first m + n – 1 constraints of the transportation problem:

$$\sum_{i=1}^{m} xij = bj \text{ and } \sum_{j=1}^{n} xij = a_i \ (j = 1, 2, \ldots n-1,\ i = 1, 2, \ldots m)$$

Taking summation on both sides, these yield

$$\sum_{j=1}^{n-1}\sum_{j=1}^{m} xij = \sum_{j=1}^{n-1} bj \text{ and } \sum_{i=1}^{m}\sum_{j=1}^{n} xij = \sum_{i=1}^{m} aj$$

Theorem:

(Existence of feasible solution): A necessary and sufficient condition for the existence of a feasible solution to the transportation problem (1) is that

$$\sum_{i=1}^{m} ai = \sum_{j=1}^{n} bj$$

Proof:

The condition is necessary: Let there exist a feasible solution to the T.P.

$$\text{Minimize } Z = \sum_{i=1}^{m}\sum_{j=1}^{n} xij\ cij$$

subject to the condition

$$\sum_{j=1}^{n} xij \quad a_i\ i = 1, 2, 3 \ldots m$$

$$\sum_{i=1}^{m} xij = b_j \qquad j = 1, 2, 3 \ldots n$$

$x_{ij} \geq 0$ for all i and j

Then we have

$$\sum_{l=1}^{m}\sum_{j=1}^{n} xij = \sum_{j=1}^{m} ai \text{ and } \sum_{j=1}^{n}\sum_{i=1}^{m} xij = \sum_{j=1}^{n} bj$$

$$\text{yielding } \sum_{i=1}^{m} ai = \sum_{j=1}^{n} bj$$

The condition is sufficient: Let us assume that

$$\sum_{i=1}^{m} ai = \sum_{j=1}^{n} bj = \lambda \text{ (say)}$$

we assert that there exist a feasible solution given by $x_{ij} = a_i b_j/\lambda$ for all i and j. Clearly $x_{ij} \geq 0$ since $a_i > 0$, $b_j > 0$ for all i and j.

Also $\sum_{j=1}^{n} xij = \sum_{j=1}^{n}(ai\,bj/\lambda) = \frac{ai}{\lambda}\cdot\left(\sum_{j=1}^{n} bj\right) = ai$, i = 1, 2 ... m and $\sum_{i=1}^{m} xij$

$= \sum_{i=1}^{m}(ai\,bj/\lambda) = b_j/\lambda\left(\sum_{j=1}^{m} ai\right) = b_j$, = j = 1, 2 ... n thus x_{ij} satisfies all the constraints of the T.P hence is a feasible solution.

THE ASSIGNMENT PROBLEM

Introduction

The assignment problem is a special case of transportation problem in which the objective is to assign a number of assigns to the equal number of destinations at a minimum cost. The assignment is to be made on one-to-one basis.

THE NATURE OF ASSIGNMENT PROBLEMS

Let there be n jobs to be performed and for doing these jobs n persons are available.

Assume that each person can do each job at a time, though with varying degree of efficiency.

Let c_{ij} be the cost (payment) of assigning ith person to the jth job. Then the problem is to find an assignment (which job should be assigned to which person) of that the total cost for performing all jobs is minimum.

The above assignment problem can be stated in the form of $n \times n$ matrix $[c_{ij}]$ of real numbers called cost *matrix* or effectiveness *matrix* as follows:

Cost Matrix

		Jobs						
		1	2	3	...	j	...	n
	1	c_{11}	c_{12}	c_{13}	...	c_{1j}	...	c_{1n}
	2	c_{21}	c_{22}	c_{23}	...	c_{2j}	...	c_{2n}
	3	c_{31}	c_{32}	c_{33}	...	c_{3j}	...	c_{3n}
	⋮	⋮	⋮	⋮		⋮		⋮
	⋮	⋮	⋮	⋮		⋮		⋮
	⋮	⋮	⋮	⋮		⋮		⋮
Persons	i	c_{i1}	c_{i2}	c_{i3}	...	c_{ij}	...	c_{in}
	⋮	⋮	⋮	⋮		⋮		⋮
	⋮	⋮	⋮	⋮		⋮		⋮
	⋮	⋮	⋮	⋮		⋮		⋮
	n	c_{n1}	c_{n2}	c_{n3}	...	c_{nj}	...	c_{nn}

MATHEMATICAL FORMULATION OF ASSIGNMENT PROBLEM

Mathematically an assignment problem can be stated as follows:
Minimize the total cost

$$Z = \sum_{i=1}^{n}\sum_{j=1}^{n} c_{ij} x_{ij}$$

where x_{ij} = {1, if ith person is assigned to the jth job 0, if not subject to the conditions

(i) $\sum_{j=1}^{n} x_{ij} = 1$, (only one job is done by the person, i = 1, 2, ... ,n)

(ii) $\sum_{i=1}^{n} x_{ij} = 1$, (only one person should be assigned to the jth job, j = 1, 2, ... ,n)

Remark:

Since x_{ij} = 0 or 1 for all i and j exactly one variable in each row or column constant |s| (and other variables is zero) the structure of the problem is an nontransportation array and such three are 2n – 1 basic variables of these exactly n has the value 1 and the remaining n – 1 the value 0. Thus, the assignment problem is heavily degenerate and it will be frustrating to attempt to solve it by transportation method.

FUNDAMENTAL THEOREMS

Now we shall prove two important theorems on which the solution to an assignment problem is fundamentally based.

Theorem:

(Reduction Theorem). *In an assignment problem if we add (or subtract) a constant to every element of a row (or column) of the cost matrix $[c_{ij}]$, then an assignment which minimizes the total cost for one matrix also minimizes the total cost for the other matrix.*

Or

Mathematical statement of reduction theorem

If $x_{ij} = X_{ij}$ *minimizes* $Z = \sum_{i=1}^{n}\sum_{j=1}^{n} c_{ij} x_{ij}$ *over all* $x_{ij} = 0$ *or 1 such that*

$\sum_{i=1}^{n} x_{ij} = 1$, $\sum_{j=1}^{n} x_{ij} = 1$, *then* $x_{ij} = X_{ij}$ *also minimizes*

$Z' = \sum_{i=1}^{n}\sum_{j=1}^{n} c_{ij}' x_{ij}$ *where* $c_{ij}' = c_{ij} \pm a_i \pm b_j$; b_i ; a_i, b_j *are some real numbers for* $i, j = 1, 2, \ldots, n$.

Proof:

We have $Z' = \sum_{i=1}^{n}\sum_{j=1}^{n} c_{ij}' x_{ij}$

$$= \sum_{i=1}^{n}\sum_{j=1}^{n} (c_{ij} \pm a_i \pm b_j) x_{ij}$$

$$= \sum_{i=1}^{n}\sum_{j=1}^{n} c_{ij} x_{ij} \pm \sum_{i=1}^{n}\sum_{j=1}^{n} a_i x_{ij} \pm \sum_{i=1}^{n}\sum_{j=1}^{n} b_j x_{ij}$$

$$= Z \pm \sum_{i=1}^{n} a_i \sum_{j=1}^{n} x_{ij} \pm \sum_{j=1}^{n} b_j \sum_{i=1}^{n} x_{ij} \qquad \left[\because Z = \sum_{i=1}^{n}\sum_{j=1}^{n} c_{ij} x_{ij}\right]$$

$$= Z \pm \sum_{i=1}^{n} a_i . 1 \pm \sum_{j=1}^{n} b_j . 1 \qquad \left[\because \sum_{i=1}^{n} x_{ij} = 1 = \sum_{j=1}^{n} x_{ij}\right]$$

$$= Z \pm \sum_{i=1}^{n} a_i \pm \sum_{j=1}^{n} b_j .$$

Since terms $\sum_{i=1}^{n} a_i$, $\sum_{j=1}^{n} b_j$ are independent of x_{ij}'s, if follows that Z' is minimized whenever Z is minimized and conversely.

Hence $x_{ij} = X_{ij}$ which minimizes Z well also minimize Z'.

Theorem 2.

If all $c_{ij} \geq 0$ *and there exists a solution* $x_{ij} = X_{ij}$ *which satisfies*

$$\sum_{i=1}^{n}\sum_{j=1}^{n} c_{ij} x_{ij} = 0,$$

then this solution is an optimal solution for the problem (i.e., minimizes the objective function).

Proof:

Since all $c_{ij} \geq 0$ and all $x_{ij} \geq 0$, the objective function $Z = \sum_{i=1}^{n}\sum_{j=1}^{n} c_{ij} x_{ij}$ cannot be negative. The minimum possible value that Z can attain is 0.

Hence the solution $x_{ij} = X_{ij}$ for which $\sum_{i=1}^{n}\sum_{j=1}^{n} c_{ij} x_{ij} = 0$ is an optimal solution.

ASSIGNMENT ALGORITHM (HUNGARIAN ASSIGNMENT METHOD)

Various steps of the computational procedure for obtaining an optimal assignment are as follows.

Step 1. Modify cost matrix by subtracting the minimum element of each row of the cost matrix, from all the elements of respective rows. Further, modify the resulting matrix by subtracting the minimum element of each column from all the elements of the respective columns. These operations create zeros.

Step 2. In the modified matrix search for an optimal assignment.

Starting with row 1 or the matrix obtained in step 1, examine rows successively until a row with exactly one zero element is found. Mark '□' at this zero as an assignment will be made there. Mark '×' at all other zeros if lying in the column containing the assigned zero. This eliminates the possibility of marking further assignments in that column. Continue in this manner until all the rows have been examined.

When the set of rows has been completely examined, an identical procedure is applied successively to columns. Starting with column 1, examine all the columns until a column containing exactly one unmarked zero is found. Then make an assignment in that position (indicated by □) and mark '×' at all zeros in the row containing this marked zero. Proceed in this way until the last column is examined.

Continue the above operations on rows and columns successively until we reach to any of the two situations

(i) all the zeros have been marked '□' or '×'

(ii) the remaining zeros lie at least two in each row and column.

In situation (i), we have a maximal assignment (assignment as much s we can) and in situation (ii) still we have some zeros to be treated. To work with such situations of zeros there is again an algorithm, complicated enough. But to avoid this highly complicated algorithm we use the trial and error method to break up such ties of zeros.

Now there are tow possibilities:

(i) If there is an assignment in every row and every column (i.e., total number of marked '□' zeros is exactly n), then we have obtained a complete optimal assignment plan.

(ii) If every row and every column do not contain an assignment (i.e., total number of marked '□' zeros is less than n), then we shall modify the cost matrix by creating some more zeros in it.

Step 3. If in step 2 every row and every column of the matrix do not contain assignment then draw the minimum number of horizontal and vertical lines to cover all the zeros at least once in the resulting matrix.

Rule to Draw Minimum Number of Lines

(i) Mark (√) all rows that do not have any marked '□' zero.

(ii) Mark (√) columns which have zeros in marked rows.

(iii) Mark (√) rows (not already marked) which have assignments in marked columns.

(iv) Repeat steps (ii) and (iii) until the chain of marking ends.

(v) Draw lines through unmarked rows and through marked columns.

This will give us the minimal system of lines.

Notes:

1. The lines thus drawn (horizontal and vertical both) are the minimum number of lines to pass through all the zeros of the matrix. It can be shown that *the minimum number of lines required to pass through all the zeros of the matrix is the same as the maximum number of assigned independent zeros of the matrix.*

Thus, if the number of lines is exactly n, then the complete assignment plan is obtained wile if the number of lines is less than n, then the complete assignment is not possible.

2. These lines cover all the zeros and each line passes through one and only one marked zero (assignment). If there are two marked '□' zeros in a row then it follow that we are assigning two jobs to one person which is a violation of the hypothesis. Thus no line passes through more than one marked '□' zero.

Step 4. Select the smallest of the elements that do not have a line though them, subtract if from all the elements that do not have a line through them, add it to every element that lies at the intersection two lines and leave the remaining elements of the matrix unchanged. In the modified matrix number of zeros are increased (never decrease) than that in step 2. Now apply the step to this new matrix. If still a complete optimal assignment is not possible in this matrix, then repeat steps 3 and 4 iteratively. Continue the process until minimum number of lines be n.

Thus exactly one marked '□' zero in each row and each column of the matrix is obtained. The assignment corresponding to these marked '□'

zero in each column of the matrix is obtained. The assignment corresponding to these marked '□' zeros will give the optimal assignment.

Note: The procedure of subtracting the minimum element of all uncovered elements from all such elements and adding this minimum element to the elements placed at the intersection does not change the optimum solution i.e., the two matrices will have the same optimal solution. For the above mentioned tow operations of addition and subtraction are the resultant of operations of subtracting the above chosen minimum element from the uncrossed rows and adding it to all the crossed columns and such operations do not change the optimum solution.

Example 1:

A car hire company has one car at each of five depots a, b, c, d and e. A customer requires a car in each town namely A, B, C, D and E. Distance (in kms) between depots (origins) and towns (destinations) are given in the following distance matrix:

	a	*b*	*c*	*d*	*d*
A	*160*	*130*	*175*	*190*	*200*
B	*135*	*120*	*130*	*160*	*175*
C	*140*	*110*	*155*	*170*	*185*
D	*50*	*50*	*80*	*80*	*110*
E	*55*	*35*	*70*	*80*	*105*

How should cars be assigned to customers so as to minimize the distance travelled?

Solution:

Step 1. Subtracting the minimum element of each row form every element of the corresponding row, the matrix reduces to

30	0	45	60	70
15	0	10	40	55
30	0	45	60	75
0	0	30	30	60
20	0	35	45	70

Now subtracting the minimum element of each column from every element of the corresponding column, the matrix reduces to

30	0	35	30	15
15	0	0	10	0
30	0	35	30	20
0	0	20	0	5
20	0	25	15	15

Step 2. Now we give the zero assignments in our usual manner,. Row 1 has a single zero in column 2. Make an assignment by marking '□' around it and delete other zeros (if any) in column 2 by marking '×', Examining the set rows completely, an identical procedure is applied successively to columns.

30	0	35	30	15
15	⊗	0	10	⊗
30	⊗	35	30	20
0	⊗	20	⊗	5
20	⊗	25	15	15

Now column 1 has a single zero in row 4. Make an assignment by marking '□' at this zero an cross the other zero of row 4 which in not yet crossed. Column 3 has a single zero in row 2, make an assignment at this zero by putting '□' and cross the other zero of row which is not yet crossed. At this stage all zeros have been either assigned or crossed out. It is observed that row 3, 5, column 4 and column 5 each has no assignment. Hence the required solution cannot be obtained at this stage. So we proceed to the next step.

Step 3. In this step we draw minimum number of lines to cover all zeros a least once. For this we proceed as follows:

(i) Mark (√) row 3 and row 5 as they have no assignments.

(ii) Mark (√) column 2 as having zeros in the marked rows 3 and 5.

(iii) Mark (√) row 1 as it contains assignment in the marked column 2.

Now further rows or colons will be required to mark during this procedure.

(iv) Now draw line L_1 through marked column 2. Then draw lines L_2 and L_3 through unmarked rows 2 and 4.

The required lines will be L_1, L_2 and L_3. No zero is left uncovered.

		L_1				
	30	[0]	35	30	15	√ (4)
L_2	15	⊗	[0]	10	⊗	
	30	⊗	35	30	20	√ (1)
L_3	[0]	⊗	20	⊗	5	
	20	⊗	25	15	15	√ (2)
		√ (3)				

Step 4. In this step we select the smallest element among all uncovered elements of the matrix of step 3.

Here this element is 15. Subtracting this element 15 from all the elements that do not have a line through them and adding to every element that lies at the intersecting of two lines and leaving the remaining elements unchanged we get the following matrix.

15	0	20	15	0
15	15	0	10	0
15	0	20	15	5
0	15	20	0	5
5	0	10	0	0

Step 5. Now again performing the step 2 we make the zero assignments. It is observed that there are no remaining zeros and every row (column) has an assignment as shown in the table.

15	⊗	20	15	[0]
15	15	[0]	10	⊗
15	[0]	20	15	5
[0]	15	20	⊗	5
5	⊗	10	[0]	⊗

Thus the complete optimal assignment plan is given by

A → e, B → c, C → b, D → a, E → d.

Form the original matrix, the minimum cost (distance travelled)

= (200 + 130 + 110 + 50 + 80) kms. = 570 kms.

Example 1:

A department head has four subordinates, and four tasks have to be performed. Subordinates differ in efficiency and tasks differ in their intrinsic difficulty. Time each man would take to perform each task is given in the electiveness matrix below. How should the tasks be allocated, one to a man, so as to minimize the total man hours?

Subordinates

		I	*II*	*III*	*IV*
	A	*8*	*26*	*17*	*11*
Tasks	*B*	*13*	*28*	*4*	*26*
	C	*38*	*19*	*18*	*15*
	D	*19*	*26*	*24*	*10*

Solution:

We shall solve this problem step by step to understand the method described above.

Step 1. Subtracting the minimum element of each row from every element of the corresponding row, the matrix reduces to

	I	II	III	IV
A	0	18	9	3
B	9	24	0	22
C	23	4	3	0
D	9	16	14	0

Now subtracting the minimum element of each column form every element of the corresponding column, the matrix reduces to

	I	II	III	IV
A	0	14	9	3
B	9	20	0	22
C	23	0	3	0
D	9	12	14	0

Step 2. Now test whether it is possible to make an assignment using only zeros.

Starting with row 1 of the matrix examine the rows one by one until a row containing exactly single zero element if found. We mark '□' at this zero, i.e., make an assignment and mark a cross '×' over all zeros if lying in the column continuing the assigned zero. Continue in this manner until all the rows have been examined. The illustration of the procedure is shown in he following table.

	I	II	III	IV
A	[0]	14	9	3
B	9	20	[0]	22
C	23	[0]	3	⊗
D	9	12	14	[0]

Now starting with column 1, examine all the columns until a column containing exactly one zero is found. We mark '□' in that position i.e., make an assignment cross other zeros if lying in the row containing this marked zero. Continue in this manner until all the columns have been examined. The illustration of the procedure is shown in the following table.

	I	II	III	IV
A	[0]	14	9	3
B	9	20	[0]	22
C	23	[0]	3	⊗
D	9	12	14	[0]

At this stage all zeros have been either assigned or crossed out. We observe that every row and every column have one assignment, so we have the complete 'zero assignment' given by

A → I, B → III, C → II, D → IV.

This assignment is also optimal for the original matrix.

The minimum total man hours = 8 + 4 + 19 + 10 = 41.

Example 4:

Solve assignment problem represented by the following matrix:

	I	*II*	*III*	*IV*	*V*	*VI*
A	*9*	*22*	*58*	*11*	*19*	*27*
B	*43*	*78*	*72*	*50*	*63*	*48*
C	*41*	*28*	*91*	*37*	*45*	*33*
D	*74*	*42*	*27*	*49*	*39*	*32*
E	*36*	*11*	*57*	*22*	*25*	*18*
F	*3*	*56*	*53*	*31*	*17*	*28*

Solution:

Step 1. Subtracting the minimum element of each row from every element of the corresponding row and then subtracting the minimum element of each column form every element of the corresponding column, the matrix reduces to

0	13	49	0	0	13
0	35	29	5	10	0
13	0	63	7	7	0
47	15	0	20	2	0
25	0	46	9	4	2
0	53	50	26	4	20

Step 2. Make the 'zero assignments' in usual manner. The illustration is shown in the following table. Since row 3 and column 5 have no assignments so we proceed to the next step.

⊗	13	49	[0]	⊗	13
⊗	35	29	5	10	[0]
13	⊗	63	7	⊗	⊗
47	15	[0]	20	2	⊗
25	[0]	46	9	4	2
[0]	53	50	26	4	20

Step 3. Draw minimum number of lines to cover all zeros at least once. For this we proceed as follows:

(i) Mark (√) row 3 as having no assignment.

(ii) Mark (√) columns 2 and 6 as having zeros in marked row 3.

(iii) Mark (√) rows 5 and 2 as having assignments in the marked columns 2 and 6.

(iv) Mark (√) column 1 (not already marked) as having zero in the marked row 2.

(v) Then mark (√) row 6 as having assignment in the marked column 1.

	L_1	L_2				L_3	
L_4	⊗	13	49	[0]	⊗	13	
	⊗	35	29	5	10	[0]	√ ⑤
	13	⊗	63	7	7	⊗	√ ①
L_5	47	15	[0]	20	2	⊗	
	25	[0]	46	9	4	2	√ ④
	[0]	53	50	26	4	20	√ ⑦
	√ ⑥	√ ②				√ ③	

Now draw lines L_1, L_2, L_3 through marked columns 1, 2, 6 respectively and L_4, L_5 through unmarked rows 1, 4 respectively. This way minimum set of five lines (5 < 6) to cover all the zeros is obtained.

Step 4. Now the smallest element among all uncovered elements is 4. Subtracting this element 4 from all the uncovered elements, adding to every element that lies at the intersection of two lines and leaving the remaining elements unchanged the matrix of step 3 reduces to the new form as shown in the table.

4	17	49	0	0	17
0	35	25	1	6	0
13	0	59	3	3	0
51	19	0	20	2	4
25	0	42	5	0	2
0	53	46	22	0	20

Step 5. Repeating the step 2, make the 'zero assignments' as shown in the following table.

Thus exactly one marked '□' zero in each row and each column of the matrix is obtained.

4	17	49	[0]	⊗	17
[0]	35	25	1	6	⊗
13	⊗	59	3	3	[0]
51	19	[0]	20	2	4
25	0	42	5	⊗	2
⊗	53	46	22	[0]	20

Thus the optimal assignment is

A → IV, B → I, C → VI, D → III, E → II, F → V.

From the original matrix, minimum cost = 11 + 43 + 33 + 27 + 11 + 17 + 142.

Note:

Another optimal solution this assignment problem is shown in the following table i.e.,

A → IV, B → VI, C → II, D → III, E → V, F → I.

4	17	49	[0]	⊗	17
⊗	35	25	1	6	[0]
13	[0]	59	3	3	⊗
51	19	[0]	20	2	4
25	⊗	42	5	[0]	2
[0]	53	46	22	⊗	20

From the original matrix, minimum cost = 11 + 48 + 28 + 27 + 25 + 3 = 142.

Example 2:

Solve the minimal assignment problem whose effectiveness matrix is

	I	*II*	*III*	*IV*
A	*2*	*3*	*4*	*5*
B	*4*	*5*	*6*	*7*
C	*7*	*8*	*9*	*8*
D	*3*	*5*	*8*	*4*

Solution:

Step 1. Subtracting the minimum element of each row from every element of the corresponding row, the matrix reduces to

0	1	2	3
0	1	2	3
0	1	2	1
0	2	5	1

Now subtracting the minimum element of each column from every element of the corresponding column, the matrix reduces to

0	0	0	2
0	0	0	2
0	0	0	0
0	1	3	0

Step 2. Now test whether it is possible to make an assignment using only zeros. Here none of the rows or columns contain exactly one zeros, Therefore, we start with row 1 searching two zeros. While examining rows successively, it is observed that row 4 has two zeros. Now, arbitrarily make an assignment (indicated by □) to one of these two zeros, say zeros say zero in the column 1 and cross other zeros in row 4 and column 1. Now we find column 1 which contains only one unmarked zero in row 3. We make assignment (indicated by □) at this zero and cross all other zeros of this row. Now again we check the rows and columns for one unmarked zero. There is no such row or column. So we start with row 1 searching two unmarked zeros, say zero of column 2, and cross other zeros of row 1 (not already crossed) and column 2. Now the second row contains only one unmarked zero in third column where we can make an assignment (indicated by □).

At this stage all zeros have been either assigned or crossed out. We observe that every row and every column have one assignment, so we have the complete 'zero assignment'.

Tables 1, 2, 3 show the necessary steps for reaching the optimal assignment.

Table 1

⊗	0	0	2
⊗	0	0	2
⊗	0	0	0
[0]	1	3	⊗

Table 2

⊗	0	0	2
⊗	0	0	⊗
⊗	⊗	⊗	[0]
[0]	1	3	⊗

Table 3

⊗	[0]	⊗	2
⊗	⊗	[0]	2
⊗	⊗	⊗	[0]
[0]	1	3	⊗

Thus we get the following optimal assignment ;

$$A \rightarrow II,\ B \rightarrow III,\ C \rightarrow IV,\ D \rightarrow I.$$

Minimum cost = 3 + 6 = 8 = 3 = 20.

Note. In this example other optimal assignments are also possible. Students must try to find them. Each has the cost 20.

Example 3:

An airline that operates seven days a week, has the time table shown below. Crews must have a minimum layover of 5 hours between fights. Obtain the pair of flights that minimizes layover time away from home. For any given pair the crew will be based at the city that results in the smaller layover.

Delhi-Jaipur

Flight No.	*Departure*	*Arrival*
1	*7.00A.M.*	*8.00A.M.*
2	*8.00A.M.*	*9.00A.M.*
3	*1.30P.M.*	*2.30P.M.*
4	*6.30P.M.*	*7.30P.M.*

Jaipur-Delhi

Flight No.	*Departure*	*Arrival*
101	*8.00A.M.*	*9.15A.M.*
102	*8.30A.M.*	*9.45A.M.*
103	*12.00Noon*	*1.15P.M.*
104	*5.30P.M.*	*6.45P.M.*

For each pair, mention the town where the crew should be based.

Solution:

First we construct the tables for layover times between the flights. Suppose we pair the flight no. 1 with flight no. 103 when crew is based at Delhi. Then the time of stay at Jaipur will be the layover time away from home. Now a plane of flight no. 1 which reaches Jaipur at 8.00 A.M., cannot fly at 12.00 Noon on the same day as minimum layover time is 5 hours. So it will depart Jaipur on the next day which will result in all layover time of 28 hours. Similarly, other layover times can be calculated.

Tables for layover times in hours

When crew based at Delhi

	101	102	103	104
1	24	24.5	28	9.5
2	23	23.5	27	8.5
3	17.5	18	21.5	27
4	12.5	13	16.5	22

When crew based at Jaipur

	101	102	103	104
1	21.75	21.25	17.75	12.25
2	22.75	22.25	18.75	13.25
3	28.25	27.75	24.25	18.75
4	9.25	8.75	5.25	23.75

To avoid the fractions we measure the layover times in terms of quarter hour (0.25 hr. or 15 minutes) as one unit of time. Thus multiplying the above tables by 4, the modified tables are as follows:

When crew based at Delhi

	101	102	103	104
1	96	98	112	38
2	92	94	108	34
3	70	72	86	108
4	50	52	66	88

When crew based at Jaipur

	101	102	103	104
1	87	85	71	49
2	91	89	75	53
3	113	111	97	75
4	37	35	21	95

As a next step we combine the above two tables, choosing that base which gives a lesser layover time for each pairing. The layover times marked with denote that crew is based at Jaipur, otherwise the crew is based at Delhi.

Minimum layover time table

	101	102	103	104
1	87*	85*	71*	38*
2	91*	89*	75*	34
3	70	72	86	75*
4	37*	35*	21*	88

Now this is a usual minimal assignment problem. Solving it by usual assignment technique, finally we get the following table.

	101	102	103	104
1	4*	0*	0*	0
2	12*	8*	8*	0
3	0	0	28	50*
4	4*	0*	0*	100

Giving the zero assignments, we the following tables:

	101	102	103	104
1	4	0*	⊗	⊗
2	12*	8*	8*	0
3	0	⊗	28	50*
4	4*	⊗*	0*	100

	101	102	103	104
1	4*	⊗	0*	⊗
2	12*	8*	8*	0
3	0	⊗	28	50*
4	4*	0*	⊗*	100

From the above tables, two optimal assignments are

(i) $(1 \to 102)^*$, $(2 \to 104)$, $(3 \to 101)$, $(4 \to 103)^*$

(ii) $(1 \to 103)^*$, $(2 \to 104)$, $(3 \to 101)$, $(4 \to 102)^*$.

In both the cases minimum layover time is 210 quarter hours i.e., 52 hours 30 minutes.

2

TRANSPORTATION PROBLEM

INTRODUCTION

The transportation problem is one of the sub-classes of the L.P.S. in which the objective is to transport various amounts of a single homogenous commodity, that one initially stored at various origin, to different destinations in such a way that the total transportation cost is a minimum. Let us illustrate by formulating a typical transportation problem involving m origins and n destinations.

TRANSPORTATION PROBLEM

The transportation problem can be described as follows.

Suppose that the factories F_i {i = 1, 2,..., m} called the origins or sources produce the non-negative quantities a_i = {i = 1, 2, – m} of product and non-negative quantities b_j (j = 1, 2,..., n) of the same product are required at other n places, called the destinations, such that the total quantity produced is equal to the total quantity required

i.e., $$\sum_{i=1}^{m} a_i = \sum_{j=1}^{n} bj \qquad ...(1)$$

Also support that C_{ij} is the cost of transportation of on unit from the ith source to the jth destination. Then the problem is to determine x_{ij}, the quantity transported from the ith source to the jth destination,

MATHEMATICAL FORMULATION OF TRANSPORTATION PROBLEM

Let there be m origins and n destinations (n may or may not be equal to m) with ith origin possessing ai units of a certain product and jth destination requiring b_j units of the same product.

Assume that the total available is equal to the total required *i.e.*, $$\sum_{i=1}^{m} a_i = \sum_{j=1}^{n} b_j \qquad ...(1)$$

Let c_{ij} be the cost of transportation of one unit product from ith origin to jth destination and x_{ij} be the quantity transported from the ith origin to the jth destination. Then the problem is to determine non-negative (≥ 0) values of xij, satisfying the availability restrictions as well as the requirement restrictions, in such a way that the total transportation cost is minimized.

i.e., find x_{ij} (≥ 0) for i = 1, 2, ... , m; j = 1, 2, ... , n which minimize

$$Z = \sum_{i=1}^{m}\sum_{j=1}^{n} c_{ij}x_{ij} \qquad ...(2)$$

such that $$\sum_{j=1}^{n} x_{ij} = a_i,\ i = 1, 2, ..., m \qquad ...(3)$$

and $$\sum_{i=1}^{m} x_{ij} = b_j,\ j = 1, 2, ..., n \qquad ...(4)$$

The equations (3) and (4) may be called the row and column equations respectively.

In such a way that the total transportation cost $\sum_{i=1}^{m}\sum_{J=1}^{n} c_{iJ}x_{iJ}$ is minimized.

The transportation problem as described above can be represented in a tabular form as follows:

Destination→ Sources↓	w_1	w_2	w_j	w_n	Capacities of the sources
F_1	c_{11}	c_{12}	c_{1J}	c_{1n}	a_1
F_2	c_{21}	c_{22}	c_{2J}	c_{2n}	a_2
F_i	c_{i1}	c_{i2}	c_{ij}	c_{in}	a_i
F_m	c_{m1}	c_{m2}	c_{mJ}	c_{mn}	a_m
Requirements→	b_1	b_2	b_J	b_n	$\sum_{i=1}^{m} a_i \sum_{J=1}^{n} b_J$.

The calculations are directly on the transportation array given below which gives the curred trial solution.

Destination→ Sources↓	w_1	w_2	w_J	w_n	Capacities of the sources
F_1	x_{11}	x_{12}	x_{1j}	w_{1n}	a_1
F_2	x_{21}	x_{22}	x_{2j}	w_{2n}	a_2
F_i	x_{ij}	x_{i2}	x_{iJ}	x_{in}	a_i
F_m	x_{m1}	x_{m2}	x_{mJ}	x_{mn}	a_m
Requirements→	b_1	b_2	b_J	b_n	$\sum_{i=1}^{m} a_i \sum_{J=1}^{n} b_J$.

The above two tables can be combned together by writing the asts CiJ within the brocket ( ), as follow:

Destination→ Sources↓	w_1	w_2	w_J	w_n	Capacities of the sources
F_1	$x_{11}(c_{11})$	$x_{12}(c_{12})$	$x_{1J}(c_{1J})$	$w_{1n}(c_{1n})$	a_1
F_2	$x_{21}(c_{21})$	$x_{22}(c_{22})$	$x_{2J}(c_{2J})$	$w_{2n}(c_{2n})$	a_2
F_i	$x_{ij}(c_{12})$	$x_{i2}(c_{12})$	$x_{iJ}(c_{iJ})$	$x_{in}(c_{in})$	a_i
F_m	$x_{m1}(c_{m1})$	$x_{m2}(c_{m2})$	$x_{mJ}(c_{mJ})$	$x_{mn}(c_{mn})$	a_m
Requirements→	b_1	b_2	b_J	b_n	$\sum_{i=1}^{m} a_i \sum_{J=1}^{n} b_J$.

RELATION BETWEEN AN ASSIGNMENT PROBLEM AND TRANSPORTATION PROBLEM

An assignment problem is a special case of the transportation problem when each origin is associated with one and only one destination. The numerical evaluations of such association are called 'effectiveness' in place of 'transportation costs'. In this case m = n, all a_i and b_j are unity and each x_{ij} is limited to one of the two values 0 and 1. In these circumstances exactly of x_{ij} can be non-zero (*i.e.*, 1), one for each origin and one for each destination.

FEASIBLE SOLUTION, BASIC FEASIBLE SOLUTION AND OPTIMUM SOLUTION OF TRANSPORTATION PROBLEM

Now we shall define few terms that are used in the transportation problem.

(i) **Feasible Solution (F.S.) :** A feasible solution to a transportation problem is a sets 0 non-negative individual allocations ($x_{ij} \geq 0$) which satisfies the row and column sum restriction [*i.e.*, equations (3) and (4) of 2].

(ii) **Basic Feasible Solution (B.F.S.) :** A feasible solution of an m by n transportation problem is said to be basic if the total number of positive allocations is equal to m + n – 1 *i.e.*, one less than the sum of the number of rows and columns.

(iii) **Optimum solution.** A feasible solution (not necessarily basic) is said to be optimum if it minimizes the total transportation cost.

EXISTENCE OF REUSABLE SOLUTION

Theorem:

A necessary and sufficient condition for the existence of feasible solution of a transportation problem is

$$\Sigma a_i = \Sigma b_j \;(i = 1, 2, \ldots ,m;\; j = 1, 2, \ldots ,n).$$

Proof:

The condition is necessary: Let there exist a feasible solution to the transportation problem. Then

$$\sum_{j=1}^{n} x_{ij} = a_i, \; i = 1, 2, \ldots ,m$$

and
$$\sum_{i=1}^{m} x_{ij} \; j = b_j = 1, 2, \ldots , n.$$

Summing over all i and j respectively, we get

$$\sum_{i=1}^{m}\sum_{j=1}^{n} x_{ij} = \sum_{i=1}^{m} a_i$$

and
$$\sum_{j=1}^{n}\sum_{i=1}^{m} x_{ij} = \sum_{j=1}^{n} b_j$$

$$\Rightarrow \quad \sum_{i=1}^{m} a_i = \sum_{j=1}^{n} b_j .$$

The condition is sufficient: Let $\sum_{i=1}^{m} a_i = \sum_{j=1}^{n} b_j = k$ (say).

If $x_{ij} = \lambda_i b_j$ for all i and j, where $\lambda_i \neq 0$ is any real number, then

$$\sum_{j=1}^{n} x_{ij} = \sum_{j=1}^{n} \lambda_i b_j = \lambda_i \sum_{j=1}^{n} b_j = k\lambda_i$$

$$\Rightarrow \qquad \lambda_i = \frac{1}{k}\sum_{j=1}^{n} x_{ij} = \frac{a_i}{k}.$$

Thus, $x_{ij} = \lambda_i b_j = \dfrac{a_i b_j}{k}$, for all i and j.

As $a_i \geq 0$, $b_j \geq 0$ so $x_{ij} \geq 0$ for all i and j.

Hence a feasible solution exists.

Note: Such a transportation problem in which $\sum a_i = \sum b_j$ is termed as balanced transportation problem. Hence a balanced transportation problem always has a F.S.

Difference between a Transportation and on Assignment Problem

An assignment problem is a special case of the transportation problem is which m = n, all the a_i and b_j are unity, and each x_{ij} is limited to one of the two value 0 and 1.

In these circumstances, exactly n of x_{ij} can be non-zero, one in each row of the array and one in each column showing that only one source (person) can be assigned to each destination (job).

Some important definitions.

1. *A feasible solution:* A feasible solution to a transportation problem is a set of non-negative individual all cations ($x_{ij} \geq 0$) which satisfies the row and column sum restrictions.
2. *Basic feasible solution:* A feasible solution of a m by n transportation problem is said to be a basic feasible solution if the total number of positive allocations x_{ij} is exactly equal to m + n – 1, *i.e.* one less than the sum of the number of rows and columns.
3. *Optimal solution:* A feasible solution (necessarily basic) is said to be optimal it minimizes the total transportation cost.
4. *Now-degenerate basic feasible solution:* A feasible solution of m by n transportation problem is said to be non degenerate basic feasible solution if.

(i) Total number of positive allocations is exactly equal to (m + n – 1).

(ii) these allocations are in independent portions.

BASIC FEASIBLE SOLUTION OF A TRANSPORTATION PROBLEM

A transportation problem is a special case of a linear programming problem. So the definition of B.F.S. is same as given earlier for L.P.P. But we find that in a transportation problem out of an unknowns there are only m + n – 1 basic variables. This happens due to redundancy in the constraints of the transportation problem. The condition $\sum a_i = \sum b_j$ can be used to reduce one constraint. This can be easily justified by proving the following theorem:

Theorem:

Out of (m + n) equations, there are only m + n – 1 independent equations in a transportation problem, m and n being the number of origins and destinations and any one equation can be dropped as the redundant equation.

Proof:

Consider m row equations and n – 1 column equations of the transportation problem as

$$\sum_{j=1}^{n} x_{ij} = a_i,\ i = 1, 2, \ldots, m \qquad \ldots(1)$$

and $$\sum_{i=1}^{m} x_{ij} = b_j,\ j = 1, 2, \ldots, n-1 \qquad \ldots(2)$$

Now adding m origin constraints given in (1), we get

$$\sum_{i=1}^{m}\sum_{j=1}^{n} x_{ij} = \sum_{i=1}^{m} a_i \qquad \ldots(3)$$

Also, adding (n – 1) destination constraints given in (2), we get

$$\sum_{j=1}^{n-1}\sum_{i=1}^{m} x_{ij} = \sum_{j=1}^{n-1} b_j \qquad \ldots(4)$$

Subtracting (4) from (3), we get

$$\sum_{i=1}^{m}\sum_{j=1}^{n} x_{ij} - \sum_{j=1}^{n-1}\sum_{i=1}^{m} x_{ij} = \sum_{i=1}^{m} a_i - \sum_{j=1}^{n-1} b_j$$

or $$\sum_{i=1}^{m}\left(\sum_{j=1}^{n} x_{ij} - \sum_{j=1}^{n-1} x_{ij}\right) = \sum_{j=1}^{n} b_j - \sum_{j=1}^{n-1} b_j \qquad [\because \sum a_i = \sum b_j]$$

or $\sum_{i=1}^{m} x_{in}$ = bn, which is the nth destination-constraint.

It follows that if m + n – 1 constraints are satisfied then the (m + n)th constraint will be automatically satisfied due to the condition $\sum a_i = \sum b_j$. Thus we have only (m + n – 1) linearly independent equations. Out of (m + n) equations, one (any) is redundant.

Hence the theorem is proved.

It indicates that B.F.S. will contain almost m + n – 1 positive variables, others being zero.

EXISTENCE OF AN OPTIMAL SOLUTION

Theorem:

There always exists an optimal solution to balanced transportation problem.

Proof:

We have $\sum_{i=1}^{m} a_i = \sum_{j=1}^{n} b_j$.

It follows that a feasible solution exists of the problem *i.e.*, $x_{ij} \geq 0$ for all i and j.

Form the constraints of the problem each $x_{ij} \leq \min(a_i, b_j)$.

Thus $0 \leq x_{ij} < \min(a_i, b_j)$ *i.e.*, the feasible region of the problem is non-empty, closed and bounded.

Hence, there exists an optimal solution.

LOOPS IN TRANSPORTATION TABLE AND THEIR PROPERTIES

Loop

Definition : *An ordered set of four or more cells is said to form a loop if it has the following properties:*

(i) *Any two adjacent cells of the set lie either in the same row or in the same column and*

(ii) *no three or more adjacent cells lie in the same row or in the same column.*

The first cell of the set will follow the last one in the set.

We get a closed path satisfying the above conditions (i) and (ii) if we join the cells of loop by horizontal and vertical line segments.

Consider two sets L = {(1, 1), (4, 1), (4, 4), (2, 4), (2, 3), (1, 3)}

and L' = {(3, 1), (3, 4), (2, 4), (2, 3), (2, 2), (1, 2), (1, 1)}

where (i, j)th cell of the transportation table is denoted by (i, j). Then it can be observed that the set L forms a loop while the set L' does not form a loop, because the three cells (2, 4), (2, 3) and (2, 2) lie in the same row. Diagrammatic illustration is given below:

Loop

(1, 1)	(1, 3)		
		(2, 3)	(2, 4)
(4, 1)			(4, 4)

Non-Loop

(1, 1)	(1, 2)		
	(2, 2)	(2, 3)	(2, 4)
	(3, 1)		(3, 4)

Properties

(i) Every loop has an even number of cells.

(ii) a feasible solution to a transportation problem is basic if and only if, the corresponding cells in the transportation table do not form a loop.

SOLUTION OF A TRANSPORTATION PROBLEM

The solution of a transportation problem consists of the following two steps:

Step 1. To find an initial basic feasible solution.

Step 2. To obtain an optimal solution by making successive improvements to initial basic feasible solution (obtained in step 1) until no further decrease in the transportation cost is possible.

METHODS TO FIND AN INITIAL BASIC FEASIBLE SOLUTION

Here we describe some simple methods to obtain the initial basic feasible solution.

Method 1: North-West Corner Rule

In this rule we have the following steps:

Step 1. Start with the cell (1, 1) at the north-west corner *i.e.*, the top-most left corner and allocate there maximum possible amount. Thus $x_{11} = \min(a_1, b_1)$.

Step 2.

(i) If $b_1 < a_1$ then $x_{11} = b_1$ and there is still some quantity available left in row 1. So move to the right hand cell (1, 2) and make the second allocation of amount $x_{12} = \min(a_1 - x_{11}, b_2)$ in the cell (1, 2).

(ii) If $b_1 > a_1$, then $x_{11} = a_1$ and there still some requirement left in column 1. So move vertically downwards to the cell (2, 1) and make the second allocation of amount $x_{21} = \min(a_1, b_1 - x_{11})$ in this cell.

(iii) If $b_1 = a_1$ then $x_{12} = 0$ or $x_{21} = 0$.

Start from the new north-west corner of the transportation table and allocate there as much as possible.

Step 3. Repeat steps 1 and 2 until all the available quantity is exhausted or all the requirement is satisfied.

The following example explains the method:

Example 1:

Find the initial basic feasible solution of the following transportation problem:

To

		W_1	W_2	W_3	*Available*
	F_1	2	7	4	5
	F_2	3	3	1	8
From	F3	5	4	7	7
	F_4	1	6	2	14
Requirement		7	9	18	34

Solution:

First we construct an empty 4 by 3 matrix complete with row and column requirements.

Start with the cell (1, 1) at the north-west corner (top-most left corner) and allocate it maximum possible amount. Thus $x_{11} = 5$ as minimum of $a_1 = 5$ and $b_1 = 7$ is 5.

To

		W_1	W_2	W_3	Available
	F_1	5 (2)			5
	F_2	2 (3)	6 (3)		8
From	F_3		3 (4)	4 (7)	7
	F_4			14 (2)	14
Requirement		7	9	18	

There is no amount left available at source 1 so in place of moving to right we move vertically downwards to the cell (2, 1) and allocate as much as possible there. The column 1 still needs the amount 2 and the amount 8 is available in row 2 so we allocate the maximum amount 2 to the cell (2, 1) *i.e.*, $x_{21} = 2$. Thus allocations for column 1 are complete. Now we move to the right of the cell (2, 1). Since the amount 6 is still available in row 2 and amount 9 is needed in column 2, so we allocate the maximum amount 6 is still available in row 2 and amount 9 is needed in column 2, so we allocate the maximum amount 6 in the cell (2, 2) *i.e.*, $x_{22} = 6$. This completes the allocations for row 2. Now we move vertically downwards to the cell (3, 2). In column 2 the amount 3 is still needed and in row 3 amount available is 7 so we allocate the maximum amount 3 in the cell (3, 2) *i.e.*, $x_{32} = 3$. Thus allocations for column 2 are complete. Now the amount 4 is still available in row 3 and amount 18 is needed in column 3, so we move to the cell (3, 3) and allocate the maximum amount 4 to this cell *i.e.*, $x_{33} = 4$. Thus, there is no amount left available at source 3. Now we move downwards to the cell (4, 3). The amount 14 is still needed in column 3 and an equal amount 14 is available in row 4, so we allocate the amount 14 to cell (4, 3) *i.e.*, $x_{43} = 14$. The resulting feasible solution is shown in the above table. Allocations in the cells are in such a way that the total the total in each row and each column is the same as shown against the respective rows and columns

Multiplying each individual allocation by its corresponding unit cost in ( ), and adding, the total cost corresponding to this feasible solution is = 5.2 + 2.3 + 6.3 + 3.4 + 4.7 + 14.2 = Rs. 102.

Note:

In this method we always move to the right or down, so no loop can be formulated by drawing horizontal and vertical lines to the allocations. Also at each step (allocation) at least one row or column is discarded from further consideration, while the last allocation discards both a row and a column simultaneously, so we cannot get more than (m + n – 1) individual positive allocations. Thus *we always get a non-degenerate basic feasible solution by the north west corner rule.*

Method 2: Lowest Cost Entry (Matrix Minima) Method

In this method we have the following steps:

Step 1. Examine the cost matrix carefully and find the lowest cost. Let it be c_{ij}. Then allocate x_{ij} as much as possible in the cell (i, j). $x_{ij} = \min(a_i, b_j)$.

Step 2.

(i) If $x_{ij} = a_i$, then the capacity of the ith origin is completely exhausted. In this case cross out the ith row of the transportation table and decrease the requirement b_j by a_i. Now go to step 3.

(ii) If $x_{ij} = b_j$, then the requirement of jth destination is completely satisfied. In this case cross out the jth column of the transportation table and decrease a_i by b_j. Now go to step 3.

(iii) If $x_{ij} = a_i = b_j$, then either cross-out the ith row or jth column but not both. Now go to step 3.

Step 3. Repeat steps 1 and 2 for the reduced transportation table until all the available is exhausted or all the requirement is satisfied.

Note:

If the cell of lowest cost is not unique. We can select any one of these cells.

The method is well explained by taking the same numerical example as in method 1.

		To			
		W_1	W_2	W_3	
From	F_1	(2)	2(7)	3(4)	5
	F_2	(3)	(3)	8(1)	8
	F_3	(5)	7(4)	(7)	7
	F_4	7(1)	(6)	7(2)	14
		7	9	18	

First we write the cost and requirement matrix. We examine the cost matrix and find that there is lowest cost 1 in cell (2, 3) and in (4, 1). We choose any one of these. Say the cell (2, 3) and allocate the maximum possible amount 8 to this cell. This exhausts the availability from F_2. Now leaving the second row in the reduced transportation table we find that there is lowest cost 1 in cell (4, 1). Here we allocate the maximum possible amount 7. This satisfies the requirement of W_1. Leaving the first column, in the reduced transportation table we find the lowest cost 2 in the cell (4, 3). We allocate the maximum possible amount 7 to this cell. This exhausts the availability from F_4. Leaving the 4th row, in the reduced transportation table, we find that there is lowest cost 4 in cell (1, 3) and in cell (3, 2). We allocate the amount 3 in the cell (1, 3). This satisfies the requirement of W_3. In order to satisfy the availability of F_1 we allocate the amount 2 to the cell (1, 2).

To complete the requirement of 9 units in column 2, we allocate the amount 7 to the cell (3, 2).

Thus we get the require B.F.S. shown in the above table.

The transportation cost = 2.7 + 3.4 + 8.1 + 7.4 + 7.1 + 72 = Rs. 83.

Method 3: Unit Cost Penalty Method (Vogel's Approximation Method).

In this method we have the following steps:

Step 1. Identify the smallest and next to smallest costs for each row of the transportation table. Find the difference between them for each row. Write these differences alongside the transportation table against the respective rows by enclosing them in parentheses. Write the similar differences for each column below the corresponding column. These are called 'penalties'.

Step 2. Now select the row or column for which the penalty is the largest. If a tie occurs, use any arbitrary tie breaking choice. Allocate the maximum possible amount to the cell with lowest cost in that particular row or column. Let the largest penalty correspond to ith row and let c_{ij} be the smallest cost in the ith row. Allocate the amount x_{ij} = min (a_i, b_j) in the cell (i, j).

Then we cross out the ith row or the jth column in the usual manner and construct the reduced matrix.

Step 3. Now compute the row and column penalties for the reduced transportation table and repeat the step 2. We continue this process until all the available quantity is exhausted or all the requirement is satisfied.

The method is well explained by taking the same numerical example as in method 1.

First we write the cost and requirement matrix and compute the penalties as follows:

	W_1	W_2	W_3	Available	Penalties
F_1	5(2)	(7)	(4)	5	(2)
F_2	(3)	(3)	(1)	8	(2)
F_3	(5)	(4)	(7)	7	(1)
F_4	(1)	(6)	(2)	14	(1)
Requirement	7	9	18		
Penalties	(1)	(1)	(1)		

We find that the maximum penalty (2) is associated with row 1 and row 2, so we may select any one of these. If we select row 1, then we allocate the maximum possible amount to the lowest cost cell in this row *i.e.*, cell (1, 1). Thus, x_{11} = min (5, 7) = 5. This exhausts the availability from F_1. So we cross the row 1. Leaving this row, the reduced cost and requirement matrix is as follows:

	W_1	W_2	W_3	Available	Penalties
F_2	(3)	(3)	8(1)	8	(2)
F_3	(5)	(4)	(7)	7	(1)
F_4	(1)	(6)	(2)	14	(1)
Requirement	2	9	18		
Penalties	(2)	(1)	(1)		

We note that the amount still needed to column 1 is 2.

Since the maximum penalty (2) is associated with row 1 and column 1 so we may select any one of these.

We select row 1 of this table and allocate the maximum possible amount 8 to the cell with cost 1 (lowest) in this row. This exhausts the availability from F_2 and leaves the requirement 10 for W_3. Again leaving row of F_2, the reduced transportation table is as follows:

	W_1	W_2	W_3	Available	Penalties
F_3	(5)	(4)	(7)	7	(1)
F_4	(1)	(6)	10(2)	14	(1)
Requirement	2	9	10		
Penalties	(4)	(2)	(5)		

In this table, the maximum penalty (5) is associated with column 3 so the maximum possible amount 10 is allocated to the cell with lowest cost 2 in this column. This completes the requirement of W_3. After leaving the column corresponding to W_3 the remaining table is as follows:

	W_1	W_2	Available	Penalties
F_3	(5)	7(4)	7	(1)
F_4	2(1)	2(6)	4	(5)
Requirement	2	9		
Penalties	(4)	(2)		

In this table, the maximum penalty (5) is associated with row 2, so the maximum possible amount 2 is allocated to the cell with lowest cost 1 in this row.

The remaining amount 2 available to F_4 is allocated to the cell with cost 6. In the last to meet the requirement of W_2 the amount 7 is allocated to the cell with cost 4.

Thus we get the required B. F. S. as shown in the table.

	W_1	W_2	W_3	Available
F_1	5(2)	(7)	(4)	5
F_2	(3)	(3)	8(1)	8
F_3	(5)	7(4)	(7)	7
F_4	2(1)	2(6)	10(2)	14
Requirement	7	9	18	

The total transportation cost

$$= 5.2 + 8.1 + 7.4 + 2.1 + 2.6 + 10.\ 2 = \text{Rs. } 80.$$

Note : If in the selected row or column the minimum cost is cost not unique, then allocate in that cell in which more allocations can be made.

Although, Vogel's method takes more time as compared to other methods, but it reduces the time in reaching the optimal solution. To obtain the optimal solution the students are advised to find the initial B. F. S. by Vogel's method.

EXERCISE

1. What is Transportation Problem? Give the mathematical formulation of transportation problem.
2. Explain the difference between a transporting problem and an assignment problem.
3. If all the sources are emptied and all the destinations are filled show that $\Sigma a_i = \Sigma b_j$ is a necessary and sufficient condition for the existence of a feasible solution to a transportation problem.
4. Prove that there are only $m + n - 1$ independent equations in a transportation problem, m and n being the no. of origins and destinations respectively and that any one equation can be dropped as the redundant equation.
5. Explain the north-west corner rule for obtaining an initial basic feasible solution of a transportation problem.

6. Explain the lowest cost entry method for obtaining an initial basic feasible solution of a transportation problem.

7. Explain Vogel's Approximation Method of solving a transportation problem.

8. Use north-west corner rule to determine an initial basic feasible solution to the following transportation problem:

(i)

		To				
		I	II	III	IV	Supply
Form	A	13	11	15	20	2
	B	17	14	12	13	6
	C	18	19	15	12	7
Demand		3	3	4	5	

(ii)

		Destination				
		D_1	D_2	D_3	D_4	Supply
Origin	O_1	6	4	1	5	14
	O_2	8	9	2	7	16
	O_3	4	3	6	2	5
Demand		6	10	15	4	

9. Determine an initial B.F.S. to the following transportation problem by using the north-west corner rule:

		Destination					
		I	II	III	IV	V	Supply
Origin	A	2	11	10	3	7	4
	B	1	4	7	2	1	8
	C	3	9	4	8	12	9
Demand		3	3	4	5	6	

10. Using 'lowest cost entry method' find the initial B.F.S. of the following transportation problem:

(i) *Destinations*

		A	B	C	D	Supply
	I	1	5	3	3	34
Origins	II	3	3	1	2	15
	III	0	2	2	3	12
	IV	2	7	2	4	19
Demand	21	25	17	17		

(ii) *Destination*

		D_1	D_2	D_3	D_4	Capacity
	O_1	1	2	3	4	6
Origin	O_2	4	3	2	0	8
	O_3	0	2	2	1	10
Demand		4	6	8	6	

11. Obtain an initial B.F.S. to the following transportation problem using Vogel approximation method:

To

		I	II	III	IV	Available
	A	5	1	3	3	34
	B	3	3	5	4	15
From	C	6	4	4	3	12
	D	4	1	4	2	19
Requirement		21	25	17	17	

12. Determine an initial B.F.S. to the following transportation table using (i) matrix minima method (ii) Vogel's approximation method:

Destination

		D_1	D_2	D_3	D_4	Supply
	O_1	1	2	1	4	30
Origin	O_2	3	3	2	1	50
	O_3	4	2	5	9	20
Demand		20	40	30	10	100

13. Find the initial basic feasible solution of the following transportation problem using (i) north-west corner rule (ii) matrix minima method (iii) Vogel's approximation method:

		Warehouse				
		W_1	W_2	W_3	W_4	Capacity
	F_1	19	30	50	10	7
Factory	F_2	70	30	40	60	9
	F_3	40	8	70	20	18
Requirement		5	8	7	14	

ANSWERS

8. (i) $x_{11} = 2, x_{21} = 1, x_{22} = 3, x_{23} = 2, x_{33} = 2, x_{34} = 5.$

   (ii) $x_{11} = 6, x_{12} = 8, x_{22} = 2, x_{23} = 14, x_{33} = 1, x_{34} = 4.$

9. $x_{11} = 3, x_{12} = 1, x_{22} = 2, x_{23} = 4, x_{24} = 2, x_{34} = 3, x_{35} = 6.$

10. (i) $x_{11} = 9, x_{12} = 8, x_{14} = 17, x_{23} = 15, x_{31} = 12, x_{42} = 17, x_{43} = 2.$

    (ii) $x_{12} = 6, x_{23} = 2, x_{24} = 6, x_{31} = 4, x_{33} = 6.$

11. $x_{12} = 25, x_{13} = 9, x_{21} = 15$ $x_{33} = 8, x_{34} = 4, x_{41} = 6, x_{44} = 13.$

12. For (i) and (ii) both: $x_{11} = 20, x_{13} = 10, x_{22} = 20, x_{23} = 20, x_{24} = 10, x_{32} = 20.$

13. (i) $x_{11} = 5, x_{12} = 2, x_{22} = 6, x_{23} = 3, x_{33} = 4, x_{34} = 14.$

    (ii) $x_{14} = 7, x_{21} = 2, x_{23} = 7, x_{31} = 3, x_{32} = 8, x_{34} = 7.$

    (iii) $x_{11} = 5, x_{14} = 2, x_{23} = 7, x_{24} = 2, x_{32} = 8, x_{34} = 10.$

In other words, if a F.S. involves exactly (m + n – 1) independent individual positive allocations, then it is known as non degenerate B.F.S. otherwise it is said to be degenerate B.F.S.

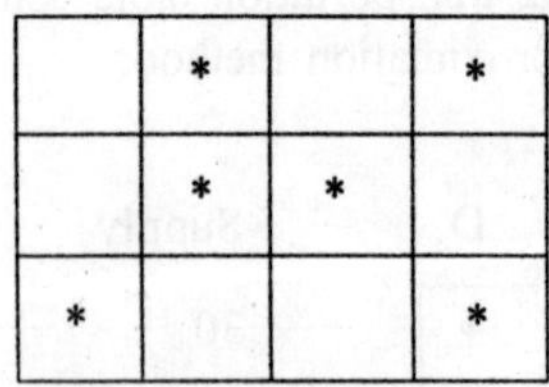

Independend poritions
(A loop is formed)

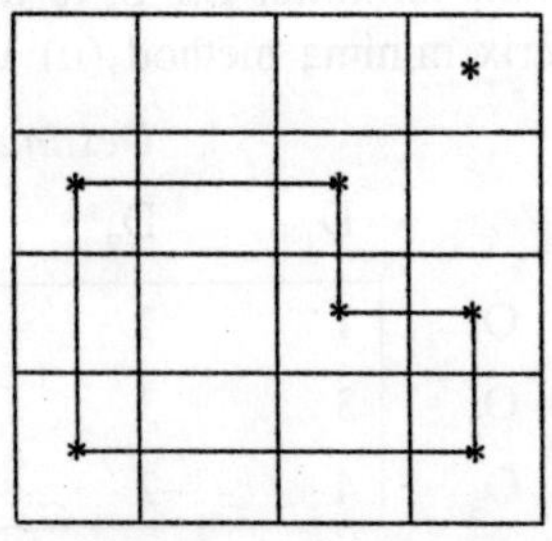

Non-Independent poition
(Donot form a loop)

Here by independent positions of the allocation we mean that it is always impossible to form any closed circuit (loop) by joining these allocation by horizontal and vertical lines only. See below the tables in which the positions of allocations are indicated by.

SOLUTION OF A TRANSPORTATION PROBLEM

The solution (optimal) of a transportation problem consists of the following two steps.

Step 1. To find on initial basic feasible solution.

Step 2. To obtain an optimal solution making improvements to initial basic passible solution until no further decrease in the transportation cot is possible.

To find an initial feasible solution.

There are several methods for finding the initial feasible solution of the given transportations problem. Here we describe the following three methods

Method 1: North-West Corner Rule

By this rule we allocate a set of allocations in the cells so that the row totals and columns totals will be as indicted before each, as follows.

(i) Start with the cell (1, 1) at the north-west corner *i.e.* the top most left corner and allocate it maximum possible amount.

(ii) Then move to the right hand cell (1, 2) if there is sill any available quantity left, dowries move to the down cell (2, 1) and allocate it maximum possible amount.

(iii) Repeat the step (ii) again and continue until all the available quantity is exhausted.

The method is well explained by toping the following numerical example:

			To		
		W_1	W_2	W_3	Supply
	F_1	(2)	(7)	(9)	5
From	F_2	(3)	(3)	(1)	8
	F_3	(5)	(4)	(7)	7
	F_4	(1)	(6)	(2)	14
Demand		7	9	18	34

To

		W_1	W_2		
	F_1	5 (2)			$5 = a_1$
From	F_2	2 (3)	6 (3)		$8 = a_2$
	F_3		3 (4)	4 (7)	$7 = a_3$
	F_4			14 (2)	$14 = a_4$
		7 b_1	$9 = b_2$	$18 = b_3$	

(i) We start with the topmost left corner and allocate is maximum possible amount 5. (Since mini of $a_1 = 5$ and $b_1 = 7$).

(ii) Since there is no amount left available at source 1 so we move downwards to the cell (2, 1), in place of moving to right, and allocate it maximum possible amount. Since the column 1 still need the amount 2 and the amount 8 is available in row 2 so we allocate the maximum amount 2 to this cell (2, 1). Now the allocation for column 1 is complete, so we move to the right of this cell. Since the amount 6 is still available in row 2 and amount 9 is needed in column 2, so we allocate the maximum amount 6 in this cell (2, Thus, allocation for row 2 is complete, so we move downwards to the cell (3, 2). Since the amount 3 is still needed in column 2 and amount 7 is available in row 3, so we allocate the maximum amount 3 in this cell (3, 2). Thus allocation for column is complete, so we move to the right of this cell. Since the amount 4 is still available in row 3 and amount 19 is needed in column 3, so we allocate the maximum amount 4 to the cell (3, 3). Thus the requirements of the row 3 is complete, so we more to the downgrade cell (4, 3). Since amount 14 is still available in row 3 and an equal amount 14 is needed in row 4, so we allocate this amount 14 to this cell (4, 3). This is complete the allocation.

In the end it may be checked that the sum of rows and columns are as needed. On multiplying each individual allocation by its corresponding unit in ( ), and adding, the total transportation cost do this F. S is

$$= 5 \times 2 + 2 \times 3 + 6 \times 3 + 3 \times 4 + 4 \times 7 + 14 \times 2$$

$$= \text{Rs. } 102.$$

In this north-west corner rule we more to the right or down, so no loop can be formulated here by drawing horizontal and vertical lines to the allocation. Also at each step at least one row or column is discarted from further consideration and at the last allocation both rows and columns are discarded. So we cannot get more than $(m + n - 1)$ individual positive allocation by this rule. Thus we always get a non degenerate basic feasible solution by this north-west corner rule.

Method 2: Lowest Cost Entry Method (Layman or Method of Matrix Minima)

In this we write the costs and the requirement matrix. The cost are written within brackets ( ). Now we examine the cost matrix carefully and choose the cell with lowest cost and allocate there as much as passible. If such cell of lowest cost in not unique, we can select any one of these cells. Again we examine the cost matrix and select the cell with the lowest cost (the cell in which allocation has been made is not considered) and allocate there as much as possible we continue this process unit all the available quantify is exhausted.

The method is well explained by taping the same number example in method 1.

First we write the cost and requirements matrix as following.

		To			
		W_1	W_2	W_3	
	F_1	(2)	2(7)	3(4)	5
From	F_2	(3)	(3)	8(1)	8
	F_3	(5)	7(4)	(7)	7
	F_4	7(1)	(6)	7(2)	14
		7	9	18	

Examine the cost matrix in above table we find there is lowest cost 1 in cell (2, 3) and in (4, 1) we choose any one of these say the cell (2, 3) and allocate the maximum amount 8 to this cell Leaving this cell we find that there is lowest best 1 in cell (4, 1) where we allocate the maximum amount 7. Continuing in this way we get the required feasible solution shows in always table.

The total transportation cost to this F. S. is

$$= 2 \times 7 + 3 \times 4 + 8 + 1 + 7 \times 4 + 7 \times 1 + 7 \times 2$$

$$= \text{Rs. } 83.$$

This cost is less than the cost associated with the F.S. obtained by north west corner rule.

The initial feasible solution obtained by this method usually gives a lower transportation cost than that obtained is north west corner rule.

Method 3: Unit Cost Penalty Method (or Vogel's Approximation Method)

In this method we write the differences of the smallest and the second smallest costs in each column, below the correspond column and write the similar differences of each row to the right of the corresponding row. These individual different can be thought of penalty for making allocations in the second lowest cost cell instead of lowest cost cell in each row or column. Now we select the row or column for which the penalty is the largest and allocate the maximum possible amount to the cell with lowest cost in that particular row or column. If there are more than one largest penalty rows or column, then we may select any one of them. Then we cross that row or column in which the requirement has been satisfied and construct the reduced matrix. We continue this process on the reduced matrices till all the allocations have been made.

The method is well explain by taking the numerical examples as in method I.

First we write the cost and requirement matrix as follows:

	W_1	W_2	W_3	Available	penalties
F_1	5(2)	(7)	(4)	5	(2)
F_2	(3)	(3)	(1)	8	(2)
F_3	(5)	(4)	(7)	7	(1)
F_4	(1)	(6)	(2)	14	(1)
Demand	7	9	18		
Penalties	(1)	(1)	(1)		

Since the maximum penalty (2) is associated with row 1 and row 2, so we may select any one of these. Suppose we select row 1 then we allocate the maximum possible amount 5 to the cell (1, 1) with lowest cost and cross this row 1.

The first reduced matrix after leaving row with remaining demand and available is and follows.

(Note that the amount 5 has been allocated to the column 1, so the amount still needed to column 1 is 2)

	W_1	W_2	W_3	Available	Penalties
F_2	(3)	(3)	8(1)	8	(2)
F_3	(5)	(4)	(7)	7	(1)
F_4	(1)	(6)	(2)	14	(1)
Demand	2	9	18		
Penalties	(2)	(1)	(1)		

Since the maximum penalty (2) is associated with row 1 above table, so the maximum possible amount 8 is allocated to the cell (1, 3) with lowest cot in this row.

The second reduced matrix after leaving row 1 of the matrix in above table with remaining demands and available is as follows

	W_1	W_2	W_3	Available	Penalties
F_3	(5)	(4)	(7)	7	(1)
F_4	(1)	(6)	10(2)	14	(1)
Demand	2	9	10		
Penalties	(4)	(2)	(5).		

Since the maximum penalty (5) is associated with column 5 of matrix in above table, so the maximum possible amount 10 is allocated to the cell (2, 3) with lowest cost in this column.

The third reduced matrix after learning column 3 of the matrix in above table with remaining demands and availables is as follows:

	W_1	W_2	Available	Penalties
F_3	(5)	7(4)	7	(1)
F_4	2(1)	2(6)	4	(5)
Demand	2	9		
Penalties	(9)	(2)		

Since the maximum penalty (5) is associated with row 2, so the maximum possible amount 2 is allocated to the cell (2, 1) with lowest cost in this row. The remaning amount 2 in row second is allocated to the cell (2, 2) and then we allocate the renaming amount 7 to the cell (1, 2) with minimum cost after crossing the row 2.

Thus we get the required feasible solution as shown in below table:

	W_1	W_2	W_3	Available
F_1	5(2)	(1)	(4)	5
F_2	(3)	(3)	8(1)	8
F_3	(5)	7(4)	(7)	7
F_4	2(1)	2(6)	10(2)	14
Demand	7	9	18	

The total transportation cost to this F.S.

$= 5 \times 2 + 8 \times 1 + 7 \times 4 + 2 \times 1 + 2 \times 6 + 10 \times 2$

= Rs. 80.

Which is less than the cost associated with the feasible solution obtained by the previous method 1 and 2.

Theorem:

It we have a B.F.S consisting of $(m + n - 1)$ independent positive allocations, and a set of arbitrary numbers u_l and v_j, $i = 1, 2, \ldots, m$, $j = 1, 2, n$, such that

$$c_{rs} = u_r + v_s$$

For all occupied cells (r, s), then the evaluation d_{ij} corresponding to each empty cell (i, j) is given by

$$d_{ij} = c_{ij} - (u_i + v_j).$$

Proof:

Mathematically a transportation problem is to determine $x_{ij} \geq 0$, which minimize.

$$Z = \sum_{i=1}^{m} \sum_{J=1}^{n} c_{iJ} \cdot x_{iJ} \qquad \text{...(1)}$$

subject to the restrictions

$$\sum_{J=1}^{n} x_{ij} = a_1 \text{ or } 0 = a_i - \sum_{J=1}^{n} x_{ij} \qquad \text{...(2)}$$

for $i = 1, 2, \ldots, m$

$$\sum_{i=1}^{m} x_{iJ} = b_j \text{ or } 0 = b_j - \sum_{i=1}^{m} x_{iJ} \qquad \text{...(3)}$$

for $j = 1, 2, \ldots, n$.

Multiplying (2) by u_i $(i = 1, 2\ldots, m)$, 3 by v_j $(= 1, 2\ldots,n)$ and adding to (11) we have

$$Z = \sum_{i=1}^{m} \sum_{J=1}^{n} c_{iJ} x_{iJ} + \sum_{i=1}^{m} u_i \left(a_i - \sum_{J=1}^{n} x_{iJ} \right) + \sum_{J=1}^{n} v_J \left(b_j - \sum_{i=1}^{m} x_{iJ} \right)$$

$$\text{or} \quad Z = \sum_{i=1}^{m} \sum_{j=1}^{n} [a_{ij} - (v_i + v_j)] x_{ij} + \sum_{i=1}^{n} v_i a_i + \sum_{j=1}^{n} v_j b_j \qquad \text{...(4)}$$

But it is given that for all occupied cells (cells with positive allocations)

$$c_{rs} = u_r + v_s \qquad \text{...(5)}$$

∴ In the objective function (4), all the terms of positive allocation vanish as this coefficients are zero.

For thus F.S the value of objective function (4), reduce to

$$Z = \sum_{i=1}^{m} u_i a_i + \sum_{j=1}^{n} v_j b_j \qquad ...(6)$$

To prove the required result, let us first determine the cell evaluation for the empty cells (i_1 j) when we allocate + 1 unit to this empty cell the number of positive allocations become (m + n), hence they become independent in position and so a closed loop can be formed. Let us closed loop formed by Joining this cell (i, j) to the occupied cells be as shown in fig (1).

Here cell (i, s), r, s) and (r, j) cre occupied cells

∴ $c_{is} = u_i + v_s$, $c_{rs} = u_r + v_s$, $c_{rj} = u_r + v_j$ when we allocate + 1 unit to the empty cell (i, j), then to maintain the row and column sums unchanged we shall decrease the individual allocation at the cell (i, s) by 1, increase that at the less (r, s) by 1 and decrease (i, j) that at cell (r, T) by 1. That the c_{ij} allocation at the occupied cells (i, s), (r, s), (r, T) are changed.

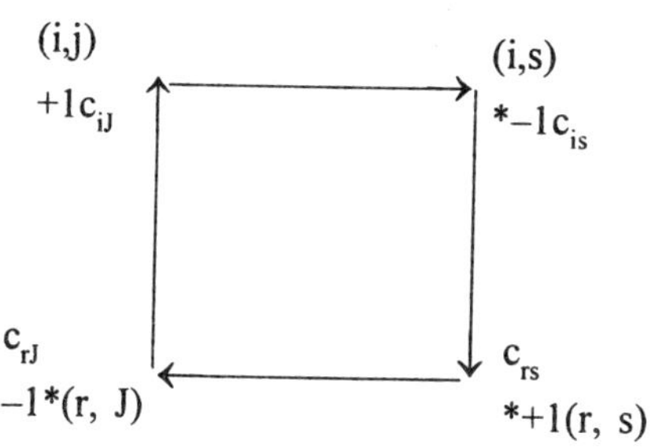

Fig. 2.1

Thus, the cell evaluation d_{ij} corresponding to this empty cell (i, j) is given by d_{ij} = Difference in the total cost for the new solution and the original one

$$= c_{iT} - c_{is} + c_{rs} - c_{rj}$$

$$= c_{ij} - (u_i + v_s) + (u_r + v_s) \;\; (u_r + v_j)$$

or $\;d_{ij} = c_{ij} - (u_i + v_j)$

This gives the evaluations d_{ij} connecting empty cell (i, j) to the occupied cells by a square shaped loop. By similar reasoning, we way generalize it to the occupied cells.

Thus, we conclude that the evaluation d_{ij} corresponding to each empty cell (i, j) is given by

$d_{ij} = c_{ij} - (u_i + v_T)$

This completes the proof of the theorem.

Computational Procedure of Optimal Test

After getting the initial B.F.S of a transportation problem, we test this solution for optometry as follows.

1. For a B.F.S. we determine a set of (m + n) numbers

$u_i \neq i = 1, 2, \ldots, m$

$v_j \neq J = 1, 2, \ldots, n$

Such that for each occupied cell (r, s)

$c_{rs} = u_r + v_s$

For this we assign an arbitrary value to one of the u_i's or v_j's then the rest (m + n – 1) of them can early resolved algebraically from the relation $c_{rs} = u_r + v_s$, for occupied cells. Generally we chooses that u_i or $v_j = 0$ for which the corresponding row or column have the maximum number of undivided allocations.

2. Then we calculate the cell evaluations d_{ij} for each unoccupied cell (i, j) by wring the formula $d_{ij} = c_{ij} - (u_i + v_j)$.

3. Then we examine the matrix of cell evaluations and crackled that

(i) If all $d_{ij} > 0$, the solution under test is optimal and unique.

(ii) If all $d_{ij} \geq 0$, with at cost one $d_{ij} = 0$, then the solution under test is optimal and alternative optimal solution exists.

(iii) If at test one $d_{ij} < 0$, then the solution is not optimal. In the last case we proceed to the next step 4.

4. If at least one $d_{ij} < 0$, in step 3, then we form a new B.F.S. In the next B.F.S we give maximum allocation to the cell for which d_{ij} is minimum and negative, by making an occupied cell empty.

5. Then we repeat the steps 1 to 3 to test the optimality to this new B.F.S.

Example 1:

Solve the following problem.

	D_1	D_2	D_3	D_4	*Supply*
O_1	*1*	*2*	*1*	*4*	*30*
O_2	*3*	*3*	*2*	*1*	*50*
O_3	*4*	*2*	*5*	*9*	*20*
Demand	*20*	*40*	*30*	*10*	*100/100.*

Solution:

Step 1. By 'Lowest Cost Entry' method an initial B.F.S of the given problem is given by the following table 1.

Table 1

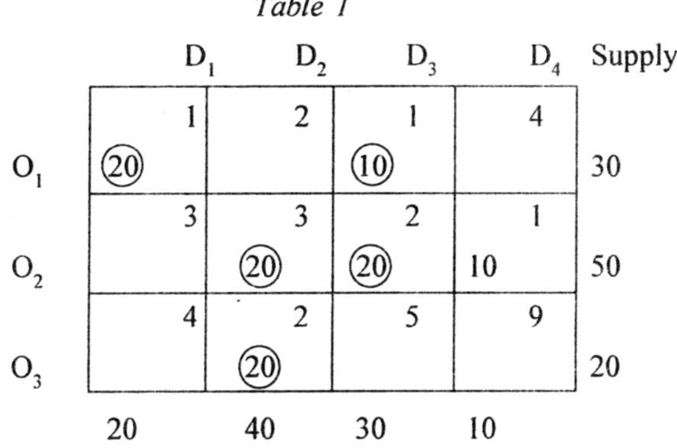

	D_1	D_2	D_3	D_4	Supply
O_1	1 (20)	2	1 (10)	4	30
O_2	3	3 (20)	2 (20)	1 10	50
O_3	4	2 (20)	5	9	20
Demand	20	40	30	10	

Total transportation cost

$= 20 \times 1 + 10 \times 1 + 20 \times 3 + 20 \times 2 + 10 \times 1 + 20 \times 2$

$= 20 + 10 + 60 + 40 + 10 + 40$

Rs. 180.

Step 2. In step 2 we will find u_i and v_j. For this we choose $u_2 = 0$ (Since row 2 contains max. number of allocations.)

$c_{22} = u_2 + v_2$ $c_{23} = u_2 + v_3$ $c_{21} = u_2 + v_4$

$3 = 0 + v_2$ $\quad 2 = 0 + v_3$ $\quad 1 = 0 + v_4$

$\Rightarrow v_2 = 3$ $\quad \Rightarrow v_3 = 2$ $\quad \Rightarrow v_4 = 1$

Also

$c_{13} = u_1 + v_3$ $\quad c_{11} = u_1 + v_1$ $\quad c_{32} = u_3 + v_2$

$1 = u_1 + 2$ $\quad 1 = -1 + v_1$ $\quad 2 = u_3 + 3$

4 $\quad u_1 = -1$ $\quad \Rightarrow v_1 = 2$ $\quad u_3 = -1.$

Step 3. Now we find the cell evaluations $u_i + v_j$ for each unoccupied cell (i, J) and enter at upper left corner of corresponding unoccupied cell.

Step 4. Then we find the cell evaluation $d_{ij} = c_{ij} - (u_i + v_j)$ (i.e., the difference of the upper right corner entry from the upper left corner entry) for each unoccupied cell (i, j) and enter at the lower left corner of the corresponding unoccupied cell.

Thus we get the following table 2.

Table 2

	D_1	D_2	D_3	D_4	Supply	u_i
O_1	1 1 (20) 0	2 2 0	1 1 (10) 0	0 4 4	30	–1
O_2	2 3 1	3 3 (20) 0	2 2 (20) 0	1 1 (10) 0	50	0
O_3	1 4	2 2 (20)	1 5	0 9	20	–1
	3	0	4	9		
Demand	20	40	30	10		
V_J	2	3	2	1		

Since all $d_{ij} \geq 0$, so the solution is optimal.

The solution of the given problem is

From Demand

O_1	to	D_1	Units	20
O_1	to	D_3	Units	10
O_2	to	D_2	Units	20
O_2	to	D_3	Units	20
O_2	to	D_4	Units	10
O_3	to	D_2	Units	20 Ans.

Example 2:

Solve the following problem.

	A_1	A_2	A_3	*Available*
B_1	*2*	*7*	*4*	*5*
B_2	*3*	*3*	*7*	*8*
B_3	*5*	*4*	*1*	*7*
B_4	*1*	*6*	*2*	*14*
Required	*7*	*9*	*18*	*34/34*

Solution:

Step 1. By Vogel's method an initial B.F.S. of the given problem is given by the following :

Table 1

	A_1	A_2	A_3	Available
B_1	2 (5)	7	4	5
B_2	3	3 (8)	7	8
B_3	5	4	1 7	7
B_4	1 (2)	6 (1)	2 (11)	14
Required	7	9	18	34/34

Total transportation cost

$= 5 \times 2 + 8 \times 3 + 7 \times 1 + 2 \times 1 + 6 \times 1 + 11 \times 2$

$= 10 + 24 + 7 + 2 + 6 + 22$

$=$ Rs. 71.

Step 2. In step 2 we will find u_i and v_j.

For this we choose $u_4 = 0$ (Since row 4 contains maximum number of allocations)

Since $c_{41} = u_4 \; v_1$ $\quad (\because c_{rs} = u_r + v_s)$

$1 = 0 + v_1$

$\Rightarrow \quad v_1 = 1$

$c_{42} = u_4 + v_2 \qquad c_{43} = u_4 + v_3$

$6 = 0 + v_2 \qquad 2 = 0 + v_3$

$v_2 = 6 \qquad v_3 = 2$

Also $\quad c_{11} = u_1 + v_1 \qquad c_{22} = u_2 + v_2 \qquad c_{33} = u_3 + v_3$

$2 = u_1 + 1 \qquad 3 = u_2 + 6 \qquad 1 = u_3 + 2$

$u_1 = 1 \qquad u_2 = -3 \qquad u_3 = -1.$

Step 3. Now we find the cell evaluations $u_i + v_j$ for each unoccupied cell (i, j) and enter at upper left corner of corresponding unoccupied cell.

Step 4. Then we find cell evaluations $d_{ij} = c_{ij} - (u_i + v_j)$ (*i.e.*, the difference of the upper right corner entry from the upper left corner entry) for each unoccupied cell (i, j) and enter at the lower left corner of the corresponding uncollected cell.

Thus we get the following table.

Table 2

	A_1	A_2	A_3	Available	ui
B_1	2 2 (5) 0	7 7 0	3 4 1	5	1
B_2	–2 3 5	3 3 (8) 0	–1 7 8	8	–3
B_3	0 5 5	5 4 –1 0	1 1 (7) 	7	–1
B_4	1 1 0	6 6 (1) 0	2 2 (11) 0	14	0
Required	7	9	18	34/84	
V_i	1	6	2		

Step 5. Since $d_{32} = -1 < 0$, so the solution is not optimal.

Step 6. Since mini d_{ij} is $d_{32} = -1$ (negative), so we given maximum allocation to this cell from an occupied cell and make necessary changes in other allocations as shown in table 3.

Table 3

(5)

(8)

+1   –1

O → (7)

↑   ↓

–1(1)←(11)+1

Step 7. The new B.F.S is shown table 4. For this B.F.S total transportation cost

$= 5 \times 2 + 8 \times 3 + 1 \times 4 + 6 \times 1 + 12 \times 2$

$= 10 + 24 + 4 + 6 + 24$

= Rs. 68

which is less than initial B.F.S.

Table 4

2 ⑤	7	4
3	3 ⑧	7
5	4 ①	1 ⑥
②	6	2 ⑫

Step 8. Proceeding as in step 2, 3 and 4 we get the following table.

Table 5

	A_1	A_2	A_3	Available	ui
B_1	2 2 ⑤ 0	6 7 1	3 4 1	5	1
B_2	–1 3 4	3 3 ⑧ 0	0 7 7	8	–2
B_3	0 5 5	4 4 ①	1 1 ⑥	7	–1
B_4	1 1 ②	5 6 ⊗	2 2 ⑫	14	0
Required	7	9	18		
V_J	1	5	2		

Since all $d_{ij} > 0$. Hence the B.F.S is optimal which is also unique.

Hence total transportation cost

$= 5 \times 2 + 8 \times 3 + 4 \times 1 + 6 \times 1 + 12 \times 2$

$= 10 + 24 + 4 + 6 + 24$

= Rs. 68.

Thus the solution of the given transportation problem is.

From source

B_1	to	A_1	Units	5
B_2	to	A_2	Units	8
B_3	to	A_2	Units	1
B_3	to	A_3	Units	6
B_4	to	A_1	Units	2
B_2	Units	A_3	to	4 Ans.

Example 3:

Solve the following transportation problem

Plant	*A*	*B*	*C*	*D*	*Available*
X	*10*	*22*	*10*	*20*	*8*
Y	*15*	*20*	*12*	*8*	*13*
Z	*20*	*12*	*10*	*15*	*11*
Required	*5*	*11*	*8*	*8*	*32/32*

Solution:

Step 1. By Vogel's an initial B.F.S. of the given problem is given by following table 1.

Table 1

	A	B	C	D	Available
X	10 ⑤	22	10 ③	20	8
Y	15	20	12 ⑤	8 ⑧	13
Z	20	12 ⑪	10 ∈	15	11
Required	5	11	8	8	

Since the total number of allocation is 5 which is one less than m + n − 1 = 6.

Hence this solution is a degenerate solution. Now to resolve this degeneracy we allocate a very small amount ∈ to the cell (3, 3) getting 6 allocation at independent position.

Total transportation cost

$= 5 \times 10 + 3 \times 10 + 5 \times 12 + 8 \times 8 + 11 \times 12$

$= 50 + 30 + 60 + 64 + 132 =$ Rs. 336.

Step 2. In step 2 we will find u_i and v_j. For this we choose $u_2 = 0$.

$c_{23} = u_2 + v_3$; $\quad c_{24} = u_2 + v_4$

$12 = 0 + v_3$; $\quad 8 = 0 + v_4$

$\Rightarrow v_3 = 12$; $\quad \Rightarrow v_4 = 8$

Also $\quad c_{13} = u_1 + v_3 \quad c_{11} = u_1 + v_1 \quad c_{33} = u_3 + v_3$

$10 = u_1 + 12 \quad 10 = -2 + v_1 \quad 10 = u_3 + 12$

$\Rightarrow u_1 = -2 \quad v_1 = 12 \quad \Rightarrow u_3 = -2$

$c_{32} = u_3 + v_2 \quad 12 = -2 + v_2 \quad \Rightarrow v_2 = 14.$

Step 3. Now we find the cell evaluations $u_i + v_j$ for each unoccupied cell (i, j) and enter at the upper left corner of the corresponding unoccupied cell.

Step 4. Then we find the cell evaluation $d_{ij} = c_{ij} - (u_i + v_j)$ (*i.e.*, the difference of the upper right) for each unoccupied cell (i, j) and enter at the lowest left corner of the corresponding unoccupied cell.

Thus we get the following table.

Table 2

	A B	C	D	Available	ui
X	10 10 ⑤ 0 10	12 22 ③ 12	−2 10 26	−6 18	20 −2
Y	12 15 ⑤ 3 6	14 8	20 0 12 13 0 12	−4 12	8
Z	10 20 ⑪ 10	12 12 0 12	−2 10 ∈ 21	−6 11	15 −2
Required	5	11	8	8	
v_j	12	14 0	−4		

Since all $d_{il} \geq 0$, so the solution is optimal.

Thus the solution of the given problem is

X	to	A	Units	5
X	to	C	Units	3
Y	to	C	Units	5
y	to	D	Units	8
Z	to	B	Units	11.

Example 4:

Determine an initial B.F.S to the following transportations problem by using the N.W. corner rule.

Destination

	A_1	B_1	C_1	D_1	E_1	
A	*2*	*11*	*10*	*3*	*7*	*4*
B	*1*	*4*	*7*	*2*	*1*	*8*
C	*3*	*9*	*4*	*8*	*12*	*9*
	3	*3*	*4*	*5*	*6*	

Solution:

By 'N–W Corner rule' method we get the following B.F.S. of the given problem.

Table 1

	A_1	B_1	C_1	D_1	E_1	Supply
A	2 ③	11 ①	10	3	7	4
B	1	4 ②	7 ④	2 ②	1	8
C	3	9	4	8 ③	12 ⑥	9
	3	3	4	5	6	

Total transportation cost = 3 × 2 + 1 × 11 + 2 × 4 + 4 × 7 × 2 × 2 + 3 × 8 + 6 × 12

= 6 + 11 + 8 + 28 + 4 + 24 + 72

= Rs 153

Example 5:

Solve the following transportation problem.

	A	*B*	*C*	*Available*
X	*7*	*3*	*4*	*2*
Y	*2*	*1*	*3*	*3*
Z	*3*	*4*	*6*	*5*
Demand	*4*	*1*	*5*	

Solution:

By vogel' method an initial B.F.S of the given problem is given by following table.

Table 1

	A	B	C	
X	7	3	4 ②	2
Y	2	1 ①	3 ②	3
Z	3 ④	4	6 ①	5
	4	1	5	

Total transportation cost
$= 2 \times 4 + 1 \times 1 + 2 \times 3 + 4 \times 3 + 1 \times 6 = 8 + 1 + 6 + 12 + 6 =$ Rs. 33.

Now finding the set u_i (1, 2, 3), v_j (J = 1, 2, 3) such that $c_{rs} = u_r + v_s$ for occupied cell and entering $u_i + v_j$ and d_{ij} in unoccupied cells, the table 2 giving the optimal solution as follows.

	A	B	C	Available	U_i
X	1 7 6	2 3 1	4 4 ② 0	2	–2
Y	0 2 2	1 1 ① 0	3 3 ② 0	3	–3
Z	3 3 ④ 0	4 4 0	6 6 ① 0	5	0
Demand	4	1	5		
v_j	3	4	6		

Since all $d_{ij} > 0$ therefore above solution is optimal

X	to	C	Units	2
Y	to	B	Units	1
Y	to	C	Units	2
Z	to	A	Units	4
Z	to	C	Units	1

Transportation cost = Rs. 33.

Example 6:

Obtain an initial B.F.S to the following.

	S_1	S_2	S_3	S_4	*Availability*
A	*5*	*1*	*3*	*3*	*34*
B	*3*	*3*	*5*	*4*	*15*
C	*6*	*4*	*4*	*3*	*12*
D	*4*	*1*	*4*	*2*	*19*
	21	*25*	*17*	*17*	

Solution:

By vogel's method an initial B.F.S of the problem is given by following table.

Table 1

	S_1	S_2	S_3	S_4	
A	5	1 (25)	3 (9)	3	34
B	3 (15)	3	5	4	15
C	6 (4)	4	4 (8)	3	12
D	4 (2)	1	4	2 (17)	19
	21	25	17	17	

A	to	S_2	Units 25
A	to	S_3	Units 9
B	to	S_1	Units 15
C	to	S_1	Units 4
C	to	S_3	Units 8
D	to	S_1	Units 2
D	to	S_4	Units 7

Total Transportation cost

$$= 25 \times 1 + 9 \times 3 + 15 \times 3 + 4 \times 6 + 8 \times 4 + 2 \times 4 + 17 \times 2$$

$$= 25 + 27 + 45 + 24 + 32 + 8 + 34 = \text{Rs. } 195.$$

Example 7:

Solve the following transportation problems

	D_1	D_2	D_3	*Supply*
O_1	2	4	1	40
O_2	6	3	2	50
O_3	4	5	6	20
O_4	3	2	1	30
O_5	5	2	5	10
Demand	50	60	40	

Solution:

By penalty method we get following B.F.S of the given problem.

Table 1

	D_1	D_2	D_3	Supply
O_1	2 (40)	4	1	40
O_2	6	3 (10)	2 (40)	50
O_3	4 (10)	5 (10)	6	20
O_4	3	2 (30)	1	30
O_5	5	2 (10)	5	10
Demand	50	60	40	

Total transportation cost

$$= 40 \times 2 + 10 \times 3 + 40 \times 2 + 10 \times 4 + 10 \times 5 + 30 \times 2 + 10 \times 2$$

$$= 80 + 30 + 80 + 40 + 50 + 60 + 20 = \text{Rs. } 360.$$

Now finding the set $U_i = (1, 2, 3)$, v_j ($j = 1, 2, 3, 4, 5$ such that $c_{rs} = u_r + v_s$ for occupied cells and entering $u_i + v_j$ and d_{ij} in unoccupied cells, the table 2 giving the optimal solution is as follows

Table 2

			u_i
2 2 (40) 0	3 4 1	2 1 0 −1	3
2 6 4	3 3 (10) 0 0	2 2 (40)	3
4 4 (10) 0	5 5 (10) 0 2	4 6	5
1 3 2	2 2 (30) 0	1 1 . 0	2
1 5 4	2 2 (10) 0	1 5 4	2
v_J −1	0	−1	

Since $d_{13} < 0$ so solution is not optimal.

Since mini $d_{iJ} = -$ (neg.), so we give maximum allocation to this cell from an occupied cell and make necessary changes in other allocation as shown in table 3.

Table 3

−10 (40)		+10 O
	(10) +10	(40) −10
(10) +10	(10)−10	
	(30)	
	(10)	

The new B.F.S is shown in table 4. For this B.F.S total transporting cost

$$= 30 \times 2 + 10 \times 1 + 20 \times 3 + 30 \times 2 + 20 \times 4 + 30 \times 2 + 10 \times 2$$

$$= 60 + 10 + 60 + 60 + 80 + 60 + 20 = \text{Rs. } 350.$$

Which is less than initial B.F.S.

Table 4

2 (30)		1 (10)
	3 (20)	2 (30)
4 (20)		
	2 (30)	
	2 (10)	

Proceeding as usual the fifth table as follows:

Table 5

	D_1	D_2	D_3	Supply	u_i
O_1	2 2 (30) 0 2	2 4	1 1 (10) 0	40	2
O_2	3 6 3	[illegible] 3 (20) 0	2 2 (30) 0	50	3
O_3	4 4 (20) 0	4 5 1	–3 6 9	20	4
O_4	2 3 1	2 2 (30) 0	1 1 0	30	2
O_5	2 5 3	2 2 (10) 0	1 5 4	10	2
Demand	50	60	40		
v_j	0	0	–1		

Since all d_j are positive therefore solution is optimal.

Now Transportation cost

$= 30 \times 2 + 10 \times 1 + 20 \times 3 + 30 \times 2$
$+ 20 \times 4 + 30 \times 2 + 10 \times 2$

$= 60 + 10 + 60 + 60 + 80 + 60 + 20$

= Rs. 350.

Example 8:

Determine the optimum basic feasible solution to the following transportation problem:

	D_1	D_2	D_3	D_4	a_i
O_1	5	3	6	2	19
O_2	4	7	9	1	37
O_3	3	4	7	5	34
$b_j \rightarrow$	16	18	31	25	

Solution:

Step 1. Using Vogel's method, the initial basic feasible solution is obtained as follows:

	D_1	D_2	D_3	D_4	
O_1	(5)	(3) 18	(6) 1	(2)	19
O_2	(4) 12	(7)	(9)	(1) 25	37
O_3	(3) 4	(4)	(7) 30	(5)	34
	16	18	31	25	

Total transportation cost = 18.3 + 1.6 + 12.4 + 25.1 + 4.3 – 30.7

= Rs. 355.

Step 2. Now u_i (i = 1, 2 3) and v_j (j = 1, 2, 3, 4) are to be determined by means of the unit cost in the respective occupied cells only. For each occupied cell (r, s). $c_{rs} = r_r - v_s$.

Since all rows contain the same (maximum) number of allocations, so take any of the u_i (say u_3) equal to zero.

When $u_3 = 0$, $v_3 = 7$ (since $c_{33} = u_3 + v_3$; $c_{33} = 7$).

Similarly $\quad c_{31} = u_3 + v_1$

or $\quad 3 = 0 + v_1$ or $v_1 = 3$.

Again $\quad c_{21} = u_2 + v_1$

or $\quad 4 = u_2 + 3$

or $\quad u_2 = 1$.

In the same way, $c_{24} = 1 = u_2 + v_4$ gives $v_4 = 0$;

$c_{13} = 6 = u_1 + v_3$ gives $u_1 = -1$

and $\quad c_{12} = 3 = u_1 + v_2$ gives $v_2 = 4$.

and $\quad \sum_{i=1}^{m} x_{ij} = b_j$

or $\quad 0 = b_j - \sum_{i=1}^{m} x_{ij}, j = 1, 2, \ldots, n$ ...(3)

Multiplying (2) by u_i ($i = 1, 2, \ldots, m$). (3) by v_j ($j = 1, 2, \ldots, n$) and adding to (1). we have

$$Z = \sum_{i=1}^{m}\sum_{j=1}^{n} c_{ij}x_{ij} + \sum_{i=1}^{m} u_i\left(a_1 - \sum_{j=1}^{n} x_{ij}\right) + \sum_{j=1}^{n} v_j\left(b_j - \sum_{j=1}^{m} x_{ij}\right)$$

or $$Z = \sum_{i=1}^{m}\sum_{j=1}^{n}[c_{ij} - (u_i + v_j)]x_{ij} + \sum_{i=1}^{m} u_i a_i + \sum_{j=1}^{n} v_j b_j \qquad \text{...(4)}$$

But it is given that for each occupied cell (r, s) (cell with positive allocation)

$$c_{rs} = u_r + v_s. \qquad \text{....(5)}$$

Hence in the objective function (4), all the terms of positive allocations vanish as their coefficients are zero. Thus, for this feasible solution the value of the objective function (4) reduces to

$$Z = \sum_{i=1}^{m} u_i a_i + \sum_{j=1}^{n} v_j b_j \qquad \text{...(6)}$$

Let us determine the cell evaluation for the empty cell (h, k). When we allocate one unit to this empty cell, the positive allocations become (m + n) in number and hence they become dependent in position. So a closed loop can be formed. Let the closed loop thus formed be as shown in the following figure.

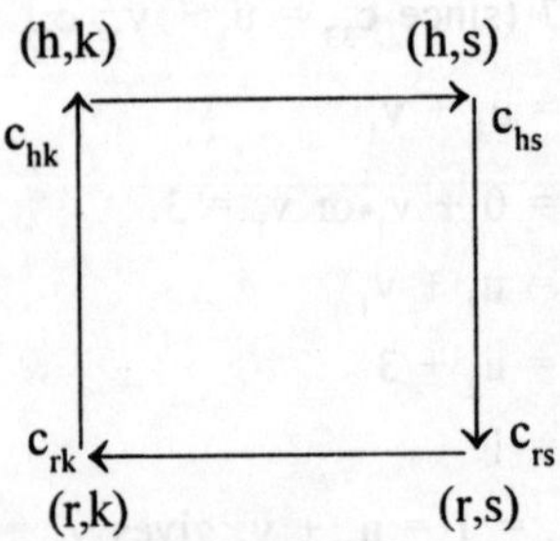

Fig. 2.2

Here cell (h, s), (r, s) and (r, k) are all occupied cells and hence

$c_{hs} = u_h + v_s$, $c_{rs} = u_r + v_s$, $c_{rk} = u_r + v_k$.

We have to decrease the individual allocations at (h, s) and (r, k) cells and increase at cell (r, s) by 1 unit to maintain the row and column sums. So the values of the individual allocations in these occupied cells are changed but in the objective function (4) they contribute nothing as their coefficients are necessarily zero. Thus, corresponding to this new solution [allocating 1 unit at cell (h, k)], the value of the objective function is given by

$$Z' = c_{hk} - (u_h + v_k) = \sum_{i=1}^{m} u_i a_i + \sum_{j=1}^{n} v_j b_j \qquad \text{...(7)}$$

Hence the cell evaluation is given by $d_{hk} = Z' - Z = c_{hk} - (u_h + v_k)$.

Thus, in general to each empty cell (i, j) the cell evaluation is given by

$$d_{ij} = c_{ij} - (u_i + v_j).$$

This completes the proof of the theorem.

Note: The above result is proved by considering a square (or rectangle) shaped loop. It can be generalized by considering a loop of an arbitrary shape connecting empty cell to the occupied cells.

Step 3. Now we find the evaluation $u_i + v_j$ for each unoccupied cell (i, j) and enter it at the upper right corner of the corresponding unoccupied cell.

Step 4. Then we find the cell evaluation $d_{ij} = c_{ij} - (u_i + v_j)$ for each unoccupied cell (i, j) and enter at the lower right corner of the corresponding unoccupied cell.

Thus we get the following table:

					$u_i \downarrow$
	(5) 2 3	(3) (6) 18	 1	(2) −1 3	−1
	(4) 12	(7) 5 2	(9) 8 1	(1) 25	1
	(3) 4	(4) 4 0	(7) 30	(5) 0 5	0
$v_j \rightarrow$	3	4	7	0	

Step 5. Since all $d_{ij} \geq 0$. The solution under test is optimal. But an alternative optimal solution will also exist as $d_{32} = 0$.

Thus the solution of the given problem is $x_{12} = 18$, $x_{13} = 1$, $x_{21} = 12$, $x_{24} = 25$, $x_{31} = 4$, $x_{33} = 30$ and minimum transportation cost = Rs. 355.

Example 9:

Find the optimum solution to the following transportation problem:

	D_1	D_2	D_3	D_4	Supply
O1	23	27	16	18	30
O2	12	17	20	51	40
O3	22	28	12	32	53
Demand	22	35	25	41	

Solution:

Step 1. By Vogel's method an initial B.F.S. of the given problem is given in the following table:

	D_1	D_2	D_3	D_4	$a_i \downarrow$
O_1	(23)	(27)	(16)	(18) 30	30
O_2	(12) 5	(17) 35	(20)	(51)	40
O_3	(22) 17	(28)	(12) 25	(32) 11	53
b_j	22	35	25	41	

Total transportation cost = 18.30 + 12.5 + 17.35 + 22.17 + 12.25 + 32.11 = Rs. 2221.

Step 2. Now we determine a set of u_i and v_j such that for each occupied cell (r, s) $c_{rs} = u_r + v_s$.

For this we choose $u_3 = 0$ (since row 3 contains maximum number of allocations).

Since $c_{31} = 22 = u_3 + v_1$, $c_{33} = 12 = u_3 + v_3$, $c_{34} = 32 = u_3 + v_4$

$\therefore v_1 = 22$, $v_3 = 12$, $v_4 = 32$.

Also $c_{21} = {}_{12} = u_2 + v_1$, $c_{22} = 17 = u_2 + v_2$, $c_{14} = 18 = u_1 + v_4$

$\therefore$ $u_2 = -10$, $v_2 = 27$, $u_1 = -14$.

Step 3. Now we find the evaluation $u_i + v_j$ for each unoccupied cell (i, j) and enter at the upper right corner of the corresponding unoccupied cell.

Step 4. Then we find the cell evaluation $d_{ij} = c_{ij} - (u_i + v_j)$ for each unoccupied cell (i, j) and enter at the lower right corner of the corresponding unoccupied cell.

Thus we get the following table:

					$u_i \downarrow$
	(23) 8 15	(27) 13 14	(16) –2 18	(18) 30	–14
	(12) 5	(17) 35	(20) 2 18	(51) 22 29	–10
	(22) 17	(28) 27 1	(12) 25	(32) 11	0
$v_j \rightarrow$	22	27	12	32	

Step 5. Since all d_{ij} for empty cell are > 0, the solution under test is optimal.

Thus the solution of the given problem is

$x_{14} = 30$, $x_{21} = 5$, $x_{22} = 35$, $x_{31} = 17$, $x_{33} = 25$, $x_{34} = 11$ and minimum transportation cost = Rs. 2221.

Example 10:

Solve the following transportation problem in which cell entries represent unit cost:

	To			Available
From	2	7	4	5
	3	3	1	8
	5	4	7	7
	1	6	2	14
Required	7	9	18	

Solution:

Step 1. Using Vogel's method, the initial B.F.S. is obtained as given below:

				a_i
	(2) 5	(7)	(4)	5
	(3)	(3)	(1) 8	8
	(5)	(4) 7	(7)	7
	(1) 2	(6) 2	(2) 10	14
$b_j \rightarrow$	7	9	18	

Total transportation cost = 2.5 + 1.8 + 4.7 + 1.2 + 6.2 + 2.10 = Rs. 80.

Step 2. To test the solution for optimality, we find a set of u_i (i = 1, 2, 3, 4) and v_j (j = 1, 2, 3) such that for each occupied cell (r, s), $c_{rs} = u_{rs} = u_r + v_s$.

For this let us choose $u_4 = 0$ (as row 4 contains maximum number of allocations).

Now $\quad c_{41} = 1 = u_4 + v_1$, $c_{42} = 6$ $u_4 + v_2$, $c_{43} = 2 = u_4 + v_3$

$\therefore \quad v_1 = 1$, $v_2 = 6$, $v_3 = 2$.

Also $\quad c_{11} = 2 = u_1 + v_1$, $c_{23} = 1 = u_2 + v_3$, $c_{32} = 4 = u_3 + v_2$

$\therefore \quad u_1 = 1$, $u_2 = -1$, $u_3 = -2$.

Step 3. Now we find the evaluation $u_i + v_j$ for each unoccupied cell (i, j) and enter it at the upper right corner of the corresponding unoccupied cell.

Step 4. Then we find the cell evaluation $d_{ij} = c_{ij} - (u_i + v_j)$ for each unoccupied cell (i, j) and enter it at the lower right corner of the corresponding unoccupied cell.

Thus, we get the following table:

				$u_i \downarrow$
	(2) 5	(7) 7 0	(4) 3 1	1
	(3) 0 3	(3) 5 −2	(1) 8	−1
	(5) −1 6	(4) 7	(7) 0 7	−2
	(1) 2	(6) 2	(2) 10	0
$v_j \rightarrow$	1	6	2	

Step 5. Form the above table, we observe that the cell evaluation $d_{22} = -2$, is negative. Therefore, the solution under test is not optimal. The solution can be improved as shown in the next step.

Step 6. Since minimum d_{ij} is $d_{22} = -2$ (negative), so we allocate (say, θ) to the cell (2, 2) as much as possible.

Now when we have decided to include one more cell to the solution the occupied cells become dependent. Therefore, we first identify the loop joining cell (2, 2) with the occupied cells.

It is easily seen by the following rule that at the most θ = 2 units can be allocated from cell (4, 2) to cell (2, 2) still satisfying the row and column total and non-negativity restrictions on the allocations.

RULE TO DETERMINE θ

The value of θ, in general, is obtained by equating to zero the of the allocations containing –θ (not + θ) at the corners of the closed loop.

Here min [8 – θ, 2 – θ] = 0 or 2 – θ = 0 or θ = 2 units.

Thus, cell (2, 2) enters the solution, while cell (4, 2) leaves the solution *i.e.*, it becomes empty. Thus, improved B.F.S. is obtained. The illustration is shown in the following tables:

5			5
	θ	8–θ	8
	7		7
2	2–θ	10+θ	14
7	9	18	

(2) 5	(7)	(4)	5
(3)	(3) 2	(1) 6	8
(5)	(4) 7	(7)	7
(1) 2	(6)	(2) 12	14
7	9	18	

Total transportation cost foris B.F.S. = 2.5 + 3.2 + 1.6 + 4.7 + 1.2 + 2.12 = Rs. 76, which is less than that for the initial B.F.S.

Now we shall test this improved B.F.S. for optimality.

Step 7. Proceeding as in steps 2, 3 and 4 we get the following table:

				$u_i \downarrow$
	(2) 5	(7) 5 2	(4) 3 1 1	
	(3) 0 3	(3) 2	(1) 6	–1
	(5) 1 4	(4) 7	(7) 2 5	0
	(1) 2	(6) 4 2	(2) 12	0
$v_j \rightarrow$	1	4	2	

Step 8. Since all emptiest evaluations $(d_{ij}) > 0$, the solution under test is optimal.

Thus the solution of the given problem is $x_{11} = 5$, $x_{22} = 2$, $x_{23} = 6$, $x_{32} = 7$, $x_{41} = 2$, $x_{43} = 12$ and minimum transportation cost = Rs. 76.

Example:

There are three parties who supply and three who require the following quantities of coal:

Party 1	*14 tons*	*consumer A*	*6 tons*
Party 2	*12 tons*	*consumer B*	*10 tons*
Party 3	*5 tons*	*consumer C*	*15 tons*

The cost matrix is as follows:

	A	*B*	*C*
1	*6*	*8*	*4*
2	*4*	*9*	*3*
3	*1*	*2*	*6*

Find the schedule of transportation policy which minimizes the cost.

Solution:

By matrix minima method the initial B.F.S. of the problem is as follows:

Destinations

		A	B	C	Available
	1	(6) 1	(8) 10	(4) 3	14
Origins	2	(4)	(9)	(3) 12	12
	3	(1) 5	(2)	(6)	5
Required		6	10	15	

Total transportations cost = 6.1 + 8.10 + 4.3 + 3.12 + 1.5 = Rs. 139.

TO TEST THE SOLUTION FOR OPTIMALITY

Finding the set of u_i (i = 1. 2, 3), v_j (j = 1, 2, 3) such that for occupied cells $c_{rs} = u_r + v_s$ and then entering evaluation $(u_i + v_j)$ and d_{ij} in the unoccupied cells, we get the following table:

				u_i
	(6) 1	(8) 10	(4) 3	0
	(4) 5 −1	(9) 7 2	(3) 12	−1
	(1) 5	(2) 3 −1	(6) −1 7	−5
$v_j \rightarrow$	6	8	4	

Since all d_{ij} are not ≥ 0, the solution test is not optimal.

First Iteration

Here two cell evaluations are negative and they are both most negative. We shall include any one of then in the solution. Let us allocate at cell (2, 1) as much as possible.

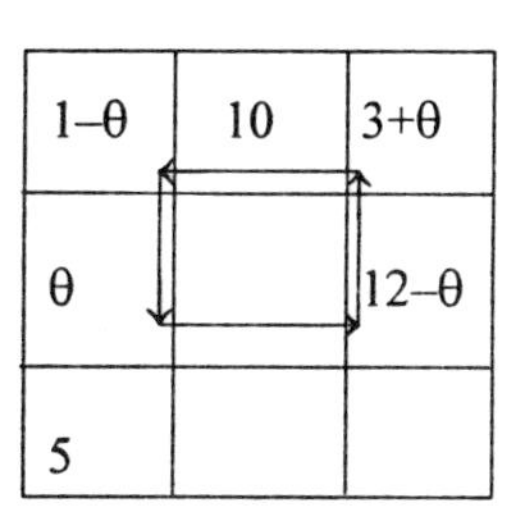

1–θ	10	3+θ
θ		12–θ
5		

Improved solution

(6)	(8) 10	(4) 4
(4) 1	(9)	(3) 11
(1) 5	(2)	(6)

Here min [1 – θ, 12 – θ] = 0 ⇒ 1 – θ = 0 ⇒ θ = 1.

Thus, cell (2, 1) enters the solution while cell (1, 1) leaves the solution *i.e.*, it becomes empty.

Transportation cost = 8.10 + 4.4 + 4.1 + 3.11 + 1.5 = Rs. 138.

To test the Improved Solution for Optimality

We have the following table giving all the necessary information:

				$u_i \downarrow$
	(6) 5 / 1	(8) 10	(4) 4	0
	(4) 1	(9) 7 / 2	(3) 11	–1
	(1) 5	(2) 4 / –2	(6) 0 / 6	–4
$v_j \rightarrow$	5	8	4	

Since all d_{ij} are not ≥ 0, the solution under test is not optimal.

Second Iteration

Since the largest negative cell evaluation is $d_{32} = -2$, so allocate as much as possible to cell (3, 2)

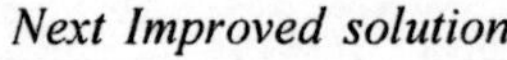

Next Improved solution

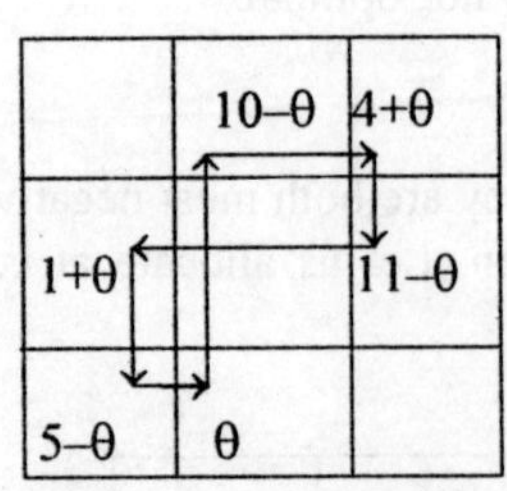

(6)	(8) 5	(4) 9
(4) 6	(9)	(3) 6
(1)	(2) 5	(6)

Here min [5 – θ, 10 – θ] = 0 5 – θ = 0 ⇒ θ = 5.

Thus, cell (3, 2) enters the solution while cell (3, 1) leaves the solution.

Transportation cost = 8.5 + 4.9 + 4.6 + 3.6 + 2.5 = Rs. 128.

To Test the Next Improved Solution for Optimality

			u_i↓
(6) 5 1	(8) 5	(4) 9	1
(4) 6	(9) 7 2	(3) 6	0
(1) –1 2	(2) 5	(6) –2 8	–5
v_j → 4	7	3	

Since all d_{ij} for empty cells are > 0, so the solution under test is optimal.

Thus the solution of the given problem is x_{12} = 5, x_{13} = 9, x_{21} = 6, x_{23} = 6, x_{32} = 5 and minimum cost = Rs. 128.

Example:

Given the following data:

		Destinations			
		1	*2*	*3*	*Capacity*
	1	*2*	*2*	*3*	*10*
Sources	*2*	*4*	*1*	*2*	*15*
	3	*1*	*3*	×	*40*
Demand		*20*	*15*	*30*	

The cost of shipment form third source to the third destination is not known. How many units should be transported from sources to the destinations so that the total cost of transporting all the units to their destinations is a minimum.

Solution:

Since the cost c_{33} is unknown, we assign a large cost, say M, to this cell. Then using Vogel's method an initial B.F.S. is obtained as shown in the table.

	1	2	3	$u_i \downarrow$
1	(2)	(2)	(3) 10	10
2	(4)	(1)	(2) 15	15
3	(1) 20	(3) 15	M 5	40
$b_j \rightarrow$	20	15	30	

To test the solution for optimality we have the following table:

			$u_i \downarrow$
(2) 4–M M–2	(2) 6–M M–4	(3) 10	3–M
(4) 3–M M+1	(1) 5–M M–4	(2) 15	2–M
(1) 20	(3) 15	(M) 5	0
$v_j \rightarrow$ 1	3	M	

Since M is very large, we observe that all the cell evaluations d_{ij} are ≥ 0. Hence, the current solution is optimum.

Thus the solution of the given problem is

$$x_{13} = 10,\ x_{23} = 15,\ x_{31} = 20,\ x_{32} = 15,\ x_{33} = 5.$$

Note: The cell (3, 3) also appears in the solution for which the cost of shipment is not known. This is known as *pseudo optimum basic feasible solution.*

DEGENERACY IN TRANSPORTATION PROBLEM

We recall that a B.F.S. to an m-origin and n-destination transportation problem will contain almost (m + n – 1) independent non-zero allocations. If this number is exactly (m + n – 1), the B.F.S. is said to be non-degenerate otherwise it is said to be a degenerate one. Thus degeneracy in transportation problem occurs whenever the number of independent individual allocations is cell than (m + n – 1).

Degeneracy in transportation problem can occur in two ways:

(i) B.F.S. may be degenerate from the initial stage onward.

(ii) It may become degenerate at any intermediate stage, when the selection of one entering cell empties two or more pre-occupied cells simultaneously.

In such cases, to resolve degeneracy, we allocate an extremely small amount (close to zero) to one or more empty cells of the matrix (generally lowest cost cells if possible), so that the total number of occupied (allocated cells become (m + n – 1) at independent positions.

The extremely small quantity usually denoted by Δ (delta) or ∈ (epsilon) satisfies the following conditions:

1. $\Delta < x_{ij}$ for $x_{ij} > 0$

2. $x_{ij} + \Delta = x_{ij} = x_{ij} - \Delta,\ x_{ij} > 0$

3. $\Delta + 0 = \Delta$.

4. If there are more than one Δ's introduced in the solution, then.

(i) if Δ, Δ' are in the same row, Δ < Δ' when Δ is to the left to Δ' and

(ii) if Δ, Δ' are in the same column, Δ < Δ' when Δ is above Δ'.

Above rules show that even after introducing Δ, the original solution of the problem is not changed. It is merely a technique to apply the optimality test. As Δ has no physical significance, ultimately it is to be omitted.

Following examples will make the procedure clear:

Example 1:

A manufacturer wants to ship 8 loads of his product as shown in the table. The matrix gives the mileage from origin O to destination D. Shipping costs are Rs. 10 per load per mile. What shipping schedule should be used?

	D_1	D_2	D_3	*Available*
O_1	50	30	220	1
O_2	90	45	170	3
O_3	250	200	50	4
Required	4	2	2	

Solution:

The initial B.F.S. by Vogel's method is obtained as follows:

	D_1	D_2	D_3	$a_i \downarrow$
O_1	(50) 1	(30)	(220)	1
O_2	(90) 3	(45)	(170)	3
O_3	(250)	(200) 2	(50) 2	4
$b_j \rightarrow$	4	2	2	

Since the total number of allocations is 4 which is one less than m + n – 1 = 5. Hence this solution is a degenerate solution. So the attempt to assign u_i and v_j values to the above table will not succeed.

To resolve this degeneracy we allocate a very small amount Δ to some suitable cell. We allocate Δ to the cell (1, 2) getting 5 allocations at independent positions.

(50) 1	(30) Δ	(220)	1+Δ = 1
(90) 3	(45)	(170)	3
(250)	(200) 2	(50) 2	4
4	2+Δ = 2	2	

To Test the Solution for Optimality

Now to test the solution for optimality we have the following table:

			$u_i \downarrow$
(50) 1	(30) Δ	(220) −120 340	−170
(90) 3	(45) 70 −25	(170) −80 250	−130
(250) 220 30	(200) 2	(50) 2	0
$v_j \rightarrow$ 220	200	50	

Since $d_{22} = -25 < 0$, so the solution under test is not optimal. Now we shall allocate to this cell (2, 2) as much as possible. Thus we take Δ from cell (1, 2) to cell (2, 2) and form the new table to check the solution for optimality.

			$u_i \downarrow$
(50) 1	(30) 5 25	(220) −145 365	−195
(90) 3	(45) Δ	(170) −105 275	−155
(250) 245 5	(200) 2	(50) 2	0
$v_j \rightarrow$ 245	200	50	

Since for empty cells all d_{ij} are > 0, so the solution under test is optimal.

Thus the solution of the given problem is

$$x_{11} = 1,\ x_{21} = 3,\ x_{32} = 2,\ x_{33} = 2$$

and minimum mileage = 50.1 + 90.3 + 200.2 + 50.2 = 820 *i.e.*, minimum cost = Rs. 8200.

Example 2:

Solve the following transportation problem:

	D_1	D_2	D_3	$a_i \downarrow$
O_1	7	4	0	5
O_2	6	8	0	15
O_3	3	9	0	9
$b_j \rightarrow$	15	6	8	

Solution:

By 'North-West Corner Rule' the non-degenerate initial B.F.S. is obtained in the following table:

	D_1	D_2	D_3	
O_1	(7) 5	(4)	(0)	5
O_2	(6) 10	(8) 5	(0)	15
O_3	(3)	(9) 1	(0) 8	9
	15	6	8	

Now to test this solution for optimality we get the following table in usual manner:

				$u_i \downarrow$
	(7) 5	(4) 9 –5	(0) 0 0	0
	(6) 10	(8) 5	(0) –1 1	–1
	(3) 7 –4	(9) 1	(0) 8	0
$v_j \rightarrow$	7	9	0	

Since all the cell evaluations are not ≥ 0, so the solution under test is not optimal.

The largest negative cell evaluation is $d_{12} = -5$, so we allocate as much as possible to the cell (1, 2).

5–θ	θ		5
10+θ	5–θ		15
	1	8	9
15	6	8	

(7)	(4) 5	(0)	5
(6) 15	(8)	(0)	15
(3)	(9) 1	(0) 8	9
15	6	8	

Here maximum value of θ is obtained by usual rule:

$\min [5 - \theta, 5, - \theta] = 0$ *i.e.*, $5 - \theta = 0$ *i.e.*, $\theta = 5$.

Since simultaneously two cells vacate, so the number of allocations becomes less than m + n – 1 *i.e.*, 5. Hence this is a degenerate solution we cannot apply the optimality test. A negligible quantity Δ may also be introduced in the independent cell (3, 1), although least cost independent cell is (2, 3).

(7)	(4) 5	(0)	5
(6) 15	(8)	(0)	15
(3) Δ	(9) 1	(0) 8	9 + Δ = 9
15 + Δ = Δ	6	8	

To test this solution for optimality we have the following table:

			u_i↓
(7) –2 9	(4) 5	(0) –5 5	–5
(6) 15	(8) 12 –4	(0) 3 –3	3
(3) Δ	(9) 1	(0) 8	0
v_j → 3	9	0	

Since all cell evaluations are not ≥ 0, so the solution under test is not optimal. The largest negative cell evaluation is $d_{22} = -4$, allocate as much as possible to the cell (2, 2).

	5	
15–θ	θ	
Δ+θ	1–θ	8

(7)	(4) 5	(0)	5
(6) 14	(8) 1	(0)	15
(3) 1	(9)	(0) 8	9
15	6	8	

Here maximum value of θ is obtained by usual rule:

$\min(15-\theta,\ 1-\theta) = 0$ *i.e.*, $1-\theta = 0$ *i.e.*, $\theta = 1$.

To test this new improved solution for optimality we have the following table:

(7) 2	(4)	(0) −1	$u_i\downarrow$
	5		−4
	5	1	
(6)	(8)	(0) 3	
14	1		0
		−3	
(3)	(9) 5	(0)	
		8	
1	4		−3
$v_j \rightarrow$ 6	8	3	

Since all the cell evaluations are not ≥ 0, so the solution under test is not optimal. The largest negative cell evaluation is $d_{23} = -3$, allocate as much as possible to the cell (2, 3).

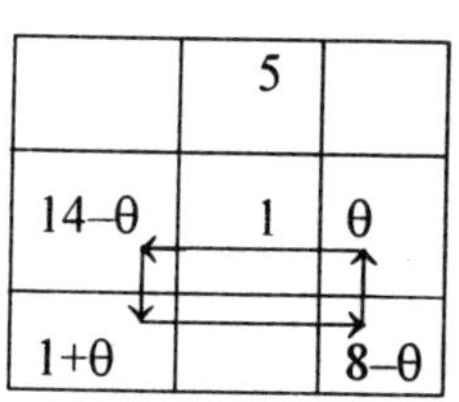

(7)	(4) 5	(0)	5
(6) 6	(8) 1	(0) 8	15
(3) 9	(9)	(0)	9
15	6	8	

Here maximum value of θ is obtained by usual rule:

$\min(14-\theta,\ 8-\theta) = 0$ *i.e.*, $8-\theta$ or $\theta = 8$.

To test this new improved solution for optimality we have the following table:

(7)	2	(4)		(0)	–4	$u_i \downarrow$
			5			–4
	5				4	
(6)		(8)		(0)		
	6		1		8	0
(3)		(9)	5	(0)	–3	
	9	8				–3
			4		3	
$v_j \rightarrow$	6		8		0	

Since all the cell evaluations for empty cells are positive, the solution under test is optimal.

Thus the solution of the given problem is

$$x_{12} = 5,\ x_{21} = 6,\ x_{22} = 1,\ x_{23} = 8,\ x_{31} = 9$$

and minimum transportation cost = 4.5 + 6 + 6 + 8.1 + 0.8 +3.9 = Rs. 91.

UNBALANCED TRANSPORTATION PROBLEM

If in a transportation problem, the sum of all available quantities is not equal to the sum all requirements i.e., if $\sum_{i=1}^{m} a_i \neq \sum_{j=1}^{n} b_j$ *then such problem is called an unbalanced transportation problem.*

An unbalanced transportation problem may occur in two different forms:

(i) *Shortage in availability, i.e.,* $\sum a_i < \sum b_j$: To modify this type of unbalanced transportation problem to balanced type we introduce a dummy source row in the transportation table. The unit transportation costs from this dummy source to any destination are all set equal to zero. The availability at this dummy source is assumed to be equal to the difference $\sum b_j - \sum a_i$.

(ii) *Excess of availability, i.e.,* $\sum a_i > \sum b_j$. To modify this type of unbalanced transportation problem to balanced type we introduce a dummy destination column in the transportation table. The unit transportation costs from this dummy destination from any source are all set equal to zero. The requirement at this dummy destination is assumed to be equal to the difference $\sum a_i - \sum b_j$.

Example:

Solve the following unbalanced transportation problem (symbols have their usual meanings).

	D_1	D_2	D_3	a_i
O_1	4	3	2	10
O_2	2	5	0	13
O_3	3	8	6	12
$b_j \rightarrow$	8	5	4	

Solution:

Here $\sum a_i = 35$, $\sum b_j = 17$. Since $\sum a_i$ is greater than $\sum b_j$, the problem is of unbalanced type. We convert this problem to a balanced one by introducing a fictitious destination D_4 with requirement 35 – 17 = 18 having all the transportation costs equal to zero. The balanced transportation table is given below:

	D_1	D_2	D_3	D_4	a_i
O_1	4	3	2	0	10
O_2	2	5	0	0	13
O_3	3	8	6	0	12
$b_j \rightarrow$	8	5	4	18	

Applying the Vogel's method in the usual manner, the initial B.F.S. is obtained as given below:

	D_1	D_2	D_3	D_4	$a_i \downarrow$
O_1	(4)	(3) 5	(2)	(0) 5	10
O_2	(2) 8	(5)	(0) 4	(0) 1	13
O_3	(3)	(8)	(6)	(0) 12	12
$b_j \rightarrow$	8	5	4	18	

This gives the transportation cost

= 3.5 + 0.5 + 2.8 + 0.4 + 0.1 + 0.12

= Rs. 31.

To test this solution for optimality we have the following table:

				$u_i \downarrow$
(4) 2 2	(3) 5	(2) 0 2	(0) 5	0
(2) 8	(5) 3 2	(0) 4	(0) 1	0
(3) 2 1	(8) 3 5	(6) 0 6	(0) 12	0
$v_j \rightarrow$ 2	3	0	0	

Since all $d_{ij} > 0$, so the solution under test is optimal.

Thus the solution of the given problem is

$$x_{12} = 5,\ x_{21} = 8,\ x_{23} = 4$$

and min. cost = Rs. 31.

EXERCISE

1. Explain how it is tested whether a B.F.S. of a transportation problem is optimal or not.
2. How does the problem of degeneracy arise in a transportation problem?
3. Explain how to solve the degeneracy in transportation problems.
4. Explain briefly the step-wise description of the computational procedure for solving the transportation problem.
5. Obtain an initial basic feasible solution to the following transportation problem. Is this solution an optimal solution? If not, obtain the optimal solution.

	W_1	W_2	W_3	W_4	a_i
F_1	19	30	50	10	7
F_2	70	30	40	60	9
F_3	40	8	70	20	18
$b_j \rightarrow$ 5	8	7	14		

6. Solve the transportation problem:

	D_1	D_2	D_3	D_4	Available
O_1	1	2	1	4	30
O_2	3	3	2	1	50
O_3	4	2	5	9	20
Required	20	40	30	10	100 Total

7. Determine the optimum basic feasible solution to the following transportation problem:

	D_1	D_2	D_3	D_4	Capacity
O_1	1	2	3	4	6
O_2	4	3	2	0	8
O_3	0	2	2	1	10
Demand 4	6	8	6		

where O_i and D_j denote ith destination respectively.

8. Is $x_{13} = 50$, $x_{14} - 20$, $x_{21} = 55$, $x_{31} = 30$, $x_{32} = 35$, $x_{34} = 25$ an optimum solution of the following transportation problem?

					Available units
	6	1	9	3	70
	11	5	2	8	55
	10	12	4	7	90
Required units	85	35	50	45	

If not, modify it to obtain an optimum basic feasible solution.

9. A company has four plants P_1, P_2, P_3, P_4 from which it supplies to three markets M_1, M_2, M_3. Determine the optimal transportation plan from the following data giving the plant to market shifting costs, quantities available at each plant and quantities required at each market.

	P_1	P_2	P_3	P_4	Required
M_1	19	14	23	11	11
M_2	15	16	12	21	13
M_3	30	25	16	39	19
Available 6	10	12	15		

10. Solve the transportation problem where all entries are unit costs.

	D_1	D_2	D_3	D_4	D_5	a_i
O_1	73	40	9	79	20	8
O_2	62	93	96	8	13	7
O_3	96	65	80	50	65	9
O_4	57	58	29	12	87	3
O_5	56	23	87	18	12	5
$b_j \rightarrow$	6	8	10	4	4	

11. The following table gives the cost for transporting material from supply points A, B C and D to demand points E, F, G, H and J.

		To				
		E	F	G	H	J
	A	8	10	12	17	15
	B	15	13	18	11	9
From	C	14	20	6	10	13
	D	13	19	7	5	12

The present allocation is as follows:

A to E 90; A to F 10; B to F 10; C to G 50; C to J 120; D to H 210; D to J 70.

Check if this allocation is optimum. If not, find an optimum schedule.

12. Solve the following transporting problem for minimum cost:

		To				
		I	II	III	IV	
	A	15	10	17	18	2
From	B	16	13	12	13	6
	C	12	17	20	11	7
		3	3	4	5	

13. Solve the following problem:

			To		Supply
	21	16	25	13	11
From	17	18	14	23	13
	32	27	18	41	19
Demand	6	10	12	15	43 Total

14. Given below the unit cost array with supplies a_i; i = 1, 2, 3 and demand b_j; j = 1, 2, 3, 4.

					$a_i \downarrow$
	8	10	7	6	50
	12	9	4	7	40
	9	11	10	8	30
$b_j \rightarrow$	25	32	40	23	

Find the optimal solution to the above problem.

15. A company has three plants A, B and C and three warehouses X, Y and Z. Number of units available at the plants is 60, 70 and 80 respectively. Demands at X, Y and Z are 50, 80 and 80 respectively. Unit costs of transportation are as follows:

	X	Y	Z
A	8	7	3
B	3	8	9
C	11	3	5

What would by your transportation plan? Give minimum distribution cost.

16. The cost-requirement table for the transportation problem is given below:

		W_1	W_2	W_3	W_4	W_5	Available
	F_1	4	3	1	2	6	40
	F_2	5	2	3	4	5	30
From							
	F_3	3	5	6	3	2	20
	F_4	2	4	4	5	3	10
Required	30	30	15	20	5		

Obtain the optimal solution of the problem.

17. Solve the following transportation problem (cell entries represent unit cost):

							Available
	5	3	7	3	8	5	3
	5	6	12	5	7	11	4
	2	1	2	4	8	2	2
	9	6	10	5	10	9	8
Required	3	3	6	2	1	2	17

19. Obtain an optimum B.F.S. to the following degenerate transportation problem:

	To			Available
From	7	3	4	2
	2	1	3	3
	3	4	6	5
Demand	4	1	5	

19. Solve the following cost minimizing transportation problem:

(i)

	D_1	D_2	D_3	D_4	Available
O_1	5	3	6	5	15
O_2	10	7	12	4	11
O_3	7	5	8	4	13
Demand	8	12	13	6	

(ii)

	D_1	D_2	D_3	D_4	D_5	Available
O_1	5	5	6	4	2	9
O_2	6	9	7	8	5	13
O_3	5	6	4	6	3	9
Demand	3	7	8	5	8	

20. Consider the following unbalanced transportation problem:

		To			
		1	2	3	Supply
From	1	5	1	7	10
	2	6	4	6	80
	3	3	2	5	15
Demand		75	20	50	

Since there is not enough supply, some of the demands at these destinations may not by satisfied. Suppose there are penalty costs for every unsatisfied demand unit which are given by 5, 3 and 2 for destination 1, 2 and 3 respectively. Find the optimal solution.

ANSWERS

5. Initial B.F.S. is $x_{11} = 5$, $x_{14} = 2$, $x_{23} = 7$, $x_{24} = 2$, $x_{32} = 8$, $x_{34} = 10$; not optimal. Optimal solution is $x_{11} = 5$, $x_{14} = 2$, $x_{22} = 2$, $x_{23} = 7$, $x_{32} = 6$, $x_{34} = 12$.

6. $x_{11} = 20$, $x_{13} = 10$, $x_{22} = 20$, $x_{23} = 20$, $x_{24} = 10$, $x_{32} = 20$; min. cost = Rs. 180.

7. $x_{12} = 6$, $x_{23} = 2$, $x_{24} = 6$, $x_{31} = 4$, $x_{33} = 6$, min. cost = Rs. 28.

8. No; $x_{12} = 30$, $x_{14} = 40$, $x_{22} = 5$, $x_{23} = 50$, $x_{31} = 85$, $x_{34} = 5$ and min. cost = Rs. 1160.

9. $x_{14} = 11$, $x_{21} = 6$, $x_{22} = 3$, $x_{24} = 4$, $x_{32} = 7$, $x_{33} = 12$; min. cost = Rs. 710.

10. $x_{13} = 8$, $x_{24} = 4$, $x_{25} = 3$, $x_{31} = 5$, $x_{32} = 4$, $x_{41} = 1$, $x_{43} = 2$, $x_{52} = 4$, $x_{55} = 1$; min. cost = Rs. 1102.

11. Not; $x_{12} = 100$, $x_{22} = 70$, $x_{25} = 80$, $x_{31} = 70$, $x_{33} = 50$, $x_{35} = 40$, $x_{44} = 210$, $x_{45} = 70$ and min. cost = Rs. 6810.

12. $x_{12} = 2$, $x_{22} = 1$, $x_{23} = 4$, $x_{24} = 1$, $x_{31} = 3$, $x_{34} = 4$; min. cost = Rs. 174.

13. $x_{14} = 11$, $x_{21} = 6$, $x_{22} = 3$, $x_{24} = 4$, $x_{32} = 7$, $x_{33} = 7$; min. cost = Rs. 796.

14. $x_{11} = 25$, $x_{12} = 2$, $x_{14} = 23$, $x_{23} = 40$, $x_{32} = 30$; min. cost = Rs. 848.

15. $x_{13} = 60$, $x_{21} = 50$, $x_{23} = 20$, $x_{32} = 80$; min. cost = Rs. 750.

16. $x_{11} = 5$, $x_{13} = 15$, $x_{14} = 20$, $x_{22} = 30$, $x_{31} = 15$, $x_{35} = 5$, $x_{41} = 10$; min. cost = Rs. 210.

17. $x_{12} = 1$, $x_{16} = 2$, $x_{21} = 3$, $x_{25} = 1$, $x_{33} = 2$, $x_{42} = 2$, $x_{43} = 4$, $x_{44} = 2$; min. cost = Rs. 103.

18. $x_{13} = 2$, $x_{22} = 1$, $x_{23} = 2$, $x_{31} = 4$, $x_{33} = 1$; min; cost Rs. 234.

19. (i) $x_{11} = 8$, $x_{12} = 7$, $x_{22} = 5$, $x_{24} = 6$, $x_{33} = 13$; min. cost = Rs. 234.

(ii) $x_{12} = 4$, $x_{14} = 5$, $x_{21} = 3$, $x_{22} = 2$, $x_{25} = 8$, $x_{32} = 1$, $x_{33} = 8$; min. cost = Rs. 154.

20. $x_{12} = 10$, $x_{21} = 60$, $x_{22} = 10$, $x_{23} = 10$, $x_{31} = 15$; min. cost = Rs. 515.

EXERCISE
(Objective Questions)

Fill in the Blanks:

Fill in the blanks "... " so that the following statements are complete and correct.

1. The transportation problem is to transport various amounts of a single homogeneous commodity, that are initially stored at various origins, to different destinations in such a way that the total transportation cost is ...

2. In balanced m × n transportation problem $\sum_{i=1}^{m} a_i$ = ..., where a_i's are capacities of the sources and b_j's are requirements of the destinations.

3. The transportation problem can be regarded as a generalization of the ....

4. A feasible solution of m by n transportation problem is said to be non-degenerate basic solution if number of positive allocations is exactly equal to....
5. By north-west corner rule we always get a ... basic feasible solution.
6. The optimality test is applicable to a F. S. consisting of ... allocations in independent positions.
7. In a transportation problem the solution under test will be optimal and unique if all the cell evaluations are....
8. In Vogel's approximation method the differences of the smallest and second smallest costs in each row and column and called....
9. In computational procedure of optimality test we choose that u_i or v_j = 0 for which the corresponding row or column has the ... number of individual allocations.
10. The iterative procedure of determining an optimum solution of a minimization transportation problem is known as ....
11. An assignment problem is a special case of an m × n transportation problem in which

    (a) m = n, (b) m = 2n, (c) 2m = n, (d) none of these.
12. In basic feasible solution of an m by n transportation problem the number of positive allocations is almost

    (a) m + n, (b) m + n – 1, (c) m – n, (d) none of these.
13. The necessary and sufficient condition for the existence of a feasible solution of a transportation problem is

    (a) $\sum a_i = \sum b_j$, (b) $\sum a_i \neq \sum b_j$, (c) $\sum a_i = 0$, (d) $\sum b_j = 0$.
14. In a transportation problem a loop may be defined as an ordered set of attest

    (a) 3 cells, (b) 4 cells, (c) 5 cells, (d) 6 cells.

15 If we have a feasible solution consisting of m + n – 1 independent allocations, and if numbers u_i and v_j satisfying $c_{rs} = u_r + v_s$, for each occupied cell (r, s) then the evaluation d_{ij} corresponding to each empty cell (i, j) is given by

(a) $d_{ij} = c_{ij} - (u_i + v_j)$ (b) $d_{ij} = c_{ij} + (u_i + v_j)$

(c) $d_{ij} = c_{ij} - (u_i - v_j)$ (d) $d_{ij} = c_{ij} + (u_i - v_j)$.

16. To improve the current B.F.S. if it is not optimal we allocate to the to cell for which d_{ij} is

(a) minimum and negative (b) maximum and positive

(c) 0 (d) none of these.

17. In a transportation problem the solution under test will be optimal if all the cell evaluations are

(a) < 0, (b) > 0, (c) ≤ 0, (d) ≥ 0.

18. In Vogel's approximation method we select the row or column for which the penalty is the

(a) largest, (b) smallest, (c) zero, (d) none of these.

19. To find an initial B.F.S. we start with the cell (1, 1) in

(a) North-West Corner Rule

(b) Lowest Cost Entry Method

(c) Vogel's approximation method

(d) none of these.

20. To find an initial B.F.S. by Matrix Minima Method, we first choose the cell with

(a) zero cost, (b) highest cost, (c) lowest cost, (d) none of these.

True of False

Write 'T' for true and 'F' for false statement.

21. A feasible solution is said to be optimal if it minimizes the total transportation cost.

22. In a non-degenerate basic feasible solution the number of positive allocations is exactly equal to $m + n - 2$.

23. We always get a non-degenerate basic feasible solution by north -west corner rule.

24. A balanced transportation problem always has a F.S.

25. There always exists a B.F.S. of a balanced transportation problem.

26. Every loop will contain an odd number of cells.

27. In optimality test for all the occupied cells, cell evaluations d_{ij} are non-zero.

28. If the cell evaluations d_{ij} are > 0 for all the occupied cells, then the optimum solution is unique.

29. There are $m + n$ independent equations in a transportation problem, m and n being the number of origins and destinations.

30. In a non-degenerate basic feasible solution the allocations must be in independent positions.

ANSWERS

1. minimum **2.** $\sum_{j=1}^{n} b_j$ **3.** assignment problem.

4. m + n – 1 **5.** non-degenerate. **6.** m + n – 1.

7. > 0. **8.** penalties. **9.** maximum.

10. Modi Method. **11.** (a). **12.** (b). **13.** (a). **14.** (b).

15. (a). **16.** (a). **17.** (d). **18.** (a). **19.** (a).

20. (c). **21.** T. **22.** F. **23.** T. **24.** T.

25. T. **26.** F. **27.** F. **28.** T. **29.** F. **30.** T.

SOLVED EXAMPLES

Example 1:

Solve the following transportation problem.

		To			
		1	*2*	*3*	*Supply*
	1	*2*	*7*	*4*	*5*
	2	*3*	*3*	*1*	*8*
From	*3*	*5*	*4*	*7*	*7*
	4	*1*	*6*	*2*	*14*
Demand		*7*	*9*	*18*	*34/34*

Solution:

By Vogel's method we get following B.F.S of the given problem.

2 (5)	7	4	5
3	3	1 (8)	8
5	4 (7)	7	7
1 (2)	6 (2)	2 (10)	14
7	9	18	

Total transportation cost = 5 × 2 + 8 × 1 + 2 × 1 + 2 × 6 + 10 × 21

= 10 + 8 + 2 + 12 + 20 + 28

= Rs. 80.

Now finding the set u_i (= 1, 2, 3, 4), v_j (= 1, 2, 3) such that $c_{rs} = u_r + v_s$ for occupied cells and entering $u_i + v_j$ and d_{ij} in unoccupied cells the table 2 giving the optimal solution is as follows :

Table 2

				u_i
	2 2 (5) 0 0	7 7 	3 4 1	1
	0 3 3	5 3 –2	1 1 (8) 0	–1
	–1 5 6	4 4 (7) 0	0 7 7	–2
	1 (2) 0	1 6 (2) 0	6 2 2 10 0	0
v_j	1	6	2	

Since $d_{22} = -2 < 0$, so the solution is not optimal.

Since minimum d_{ij} is $d_{22} = -2$ (negative), so we give maximum allocation to this cell from an occupied cell and make the necessary changes in other allocations as shown in table 3

Table 3

5	+2	–2 8
	7	+2
	–2 2	10

The new basic feasible is shown in table 4. For this B.S.F.

The total transportation cost

$= 5 \times 2 + 2 \times 1 + 2 \times 3 + 7 \times 4 + 6 \times 1 + 12 \times 2$

$=$ Rs. 76.

which is less than that for the initial B.F.S.

Table 4

				Supply
	2 5	7	4	5
	3	3 2	1 6	8
	5	4 7	7	7
	1 2	6	2 12	14
Demand	7	9	18	

Proceeding as usual the fifth table as follows:

Table 5

				u_i
	2 2 5 0	5 7 2	3 4 1	1
	0 3 3	3 3 2 0	1 1 6 0	−1
	1 5 4	4 4 7 0	2 7 5	0
	1 1 2 0	4 6 2	2 2 12 0	0
v_j	1	4	2	

Since all $d_{ij} > 0$. Here B.F.S. is an optimal solution which is also unique.

Thus, the solution of given transportation problem is

From Source 1 transport 5 units to destination 1.

"	"	2	"	2 and 6 Units	"	"	2&3.
"	"	3	"	7 Units to	"	"	2
"	"	4	"	2 and 12 Units	to	"	1&3

the total transportation cost = Rs. 76. Ans.

Example 2:

A company has four plants P_1, P_2, P_3, P_4 from which it supplies to three markets M_1, M_2, M_3. Determine the optimal transportation plan from the following data giving the plant to market shifting costs, quantities available at each plant and quantities required at each marked.

Plant

Market↓	P_1	P_2	P_3	P_4	*Required at Market*
M_1	*19*	*14*	*23*	*11*	*11*
M_2	*15*	*16*	*12*	*21*	*13*
M_3	*30*	*25*	*16*	*39*	*19*
Available at Plant	*6*	*10*	*12*	*15*	*43*

Solution:

By Vogel's method an initial B.F.S. at the given problem is given by the following table.

Table 1

	19		14		23		11	Required
						11		11
	15		16		12		21	
6		3				4		13
	30		25		16		39	
		7		12		19		
Available 6	10		12		15			

The total transportation cost

$= 6 \times 15 + 3 \times 16 + 7 \times 25 + 12 \times 16 + 11 \times 11 + 4 \times 21$

$=$ Rs. 710.

Now finding the set $u_i = (1, 2, 3)$, v_J $(J = 1, 2, 3, 4)$ such that $c_{rs} = u_r + v_s$ for occupied cells and entering $u_i + v_J$ and d_{iJ} in unoccupied cell the table 2 giving the optimal solution is as follows.

Table 2

	5		19	6		14	−3		23	11		11	u_i
											11		−10
	14			8			26		0				
	15		15	16		16	7		12	21		21	
		6			3						4		
0													
	0			0			5			0			
	24		30	25		25	16		16	30		39	
					7			12			9		
	6			0						9			
v_J			15			16			7			21	

Since all $d_{ij} > 0$. Hence B.F.S. is an optimal solution and the optimal transportation cost

= Rs. 710.

Thus, the solution of the given transportation problem is.

From source 1 transport 11 Units to destination 4.

" " 2 " 6, 3 and 4 Units to destination 1, 28 4 respectively

" " 3 7 and 12 Units to destinations 2 and 3 repetitively.

EXERCISE

1. Explain the difference between a transportation problem and an assignment problem.

2. Write a short note on transportation problem.

3. Solve the following transportation problems.

Ware house / Factory↓	W_1	W_2	W_3	W_4	Factory Capacity
F_1	10	30	50	10	7
F_2	70	30	40	60	9
F_3	40	8	70	20	18
Ware house	5	8	7	14	

Requirement

4.

	D_1	D_2	D_3	D_4	D_5	D_6	a_i
O_1	9	12	9	6	9	10	5
O_2	7	3	7	7	5	5	6
O_3	6	5	9	11	3	11	2
O_4	6	8	11	2	2	11	9
b_j	4	4	6	2	4	2	

5.

Market

Plant	A	B	C	Available at plant
X	11	21	16	14
Y	7	17	13	26
Z	11	23	21	36
Required at Market	18	28	25	

6. A manufacturer wants to ship 8 loads of his product as shown below. The matrix gives the mileage from origin 0 to the destination

O↓→	O	A	B	C Available
X	50	30	220	1
Y	90	45	170	3
Z	250	200	50	4
Required	4	2	2	8

Shipping cost are Rs. 10 per load male. What shipping schedule should be used.

7. Solve the following T.C. problems.

(i)

	A	B	C	a_i
I	10	9	8	8
II	10	7	10	7
III	11	9	7	9
IV	12	14	10	4

b_j	10	10	8

(ii)

	W_1	W_2	W_3	W_4	Capacity
F_1	95	105	80	15	12
F_2	115	180	40	30	7
F_3	155	180	95	70	5
Requirement	5	4	4	11	

8. To

		W_1	W_2	W_3	W_4	W_5	Available
	F_1	3	4	6	8	9	20
	F_2	2	10	1	5	8	30
From	F_3	7	11	20	40	3	15
	F_4	2	1	9	14	16	13
Required		40	6	8	18	6	78

9. To

		A	B	C	Available
	X	7	3	4	2
From	Y	2	1	3	3
	Z	3	4	6	5
Demand		4	1	5	10

9. Solve the following T.C problems.

	D_1	D_2	D_3	D_4	a_i
O_1	6	7	3	4	5
O_2	7	9	1	2	7
O_3	6	5	16	7	8
O_4	18	9	10	2	10
	10	5	10	5	

ANSWER

3. F_1 to W_1 Units 5

to W_4 Units 2

F_2 to W_2 Units 2

to W_3 Units 7

F_3 to W_2 Units 6

to W_4 Units 12

Total cost = Rs. 743.00.

4. O_1 to D_1 Units 5

O_2 to D_2 Units 4

to D_6 Units 2

O_3 to D_1 Units 1

to D_3 Units 1

O_4 to D_1 Units 3

to D_4 Units 2

to D_5 Units 4

Total T. cost = Rs. 112.00.

5. X to C Units 14

Y to B Units 15

to C " 11

Z to A " 18

to B " 13

5 Units are left at plant Z.

Total T. cost = Rs. 1119.00.

6. X to B Units 1

Y to A " 3

Z " B " 2

" C " 2

Total T. cost Rs. 8200.

7. (i)

I to A Units 6

to B Units 2

II " B " 7

III to B " 1

to C " 8

IVto A " 4

T. cost = Rs. 240.

(ii)

F_1	to	W_1	Units	5
	to	W_2	"	4
	to	W_4	"	3
F_3	to	W_3	"	4
	to	W_4	"	3
F_3	to	W_4	"	5

Optimal T.C = Rs. 1540.

8.

F_1	to	W_1	Units	20
F_2	to	W_1	Units	4
	to	W_3	"	8
	to	W_4	"	18
F_3	to	W_1	"	9
	to	W_5	Units	6
F_4	to	W_1	"	7
	"	W_2	"	6

9.

X	to	A	2	Units
Y	to	B	1	"
Y	to	C	2	"
Z	to	A	4	"
Z	to	C	1	"

Total T. cost Rs. 33.

③

SENSITIVITY ANALYSIS

INTRODUCTION

Following the formulation of L.P.P. and attainment of an optimum solution, it is often desired to study the effect of changes in the different parameters of the problem, on the current optimum solution. For example, suppose that an error has been discovered in the data after the attainment of an optimum solution may by due to a wrong value used for a cost coefficient or due to a variable omitted inadvertently. Clearly, there are two alternatives to a accomplish this:

(i) Solve the L.P.P. new after replacing the correct values for the wrong ones.

(ii) Discover some procedure so that the computational labour on the error-fated problem may no go waste.

Undoubtedly, one would like to go in for (ii). We shall see soon that one has just to perform certain additional computerisation, after the attainment of current optimum solution, in order to often the optimum solution to the new (post-optimal) problem. This will be the objective of the present discussion.

The problems of the type discussed above arise in general, when slight changes are made in the parameters or the structure of a given L.P.P. after its optimum solution has been attain. An analysis of such post optimal problem can thus, be formed. Post optimality analysis the changes (variations) in an L.P.P. which are usually studied by post optometry analysis indeed:

1. Changes in the requirement vector b.
2. Changes in the cost vector c these may further be classified as the changes in :

   (a) The cost coefficients associated with the basic variables, and.

(b) The cost coefficients associated with the non-basic variables.

3. Changes in the elements of coefficient matrix A, these may further be classified as the changes in.

   (a) Those column vectors of A which form a basis set, and

   (b) Those column vectors of 'A' which do not form a basis set.

4. Structural changes due to the addition and subtraction (deletion) of same variables.

5. Structural changes due to the addition and deletion of some linear constraints.

VARIATION IN THE PRICE VECTOR c

Consider the L.P.P Max. $\mathbf{Z = cx}$, **s.t.** $\mathbf{Ax = b, x \geq 0}$. If $\mathbf{x_B}$ is the optimal basic feasible solution and B the optimal basis matrix, we have

$$\mathbf{x_B = B^{-1} b}.$$

It is clear that x_B is independent of therefore the change in some component c_j of **c** will not change x_B, *i.e.*, x_B will always remain basic feasible solution.

Condition of optimality for the solution x_B, is $\Delta_j = c_j - Z_j \leq 0$ for all j not in the basis which is satisfied before any change in c_j. But when c_j changes, the condition of optometry may not be satisfied. The change in the price vector c maybe made in the following two ways:

(i) Variation in $c_j \notin c_B$

(i.e., Change in c_j Which is the Price of the Non-Basic Variable x_j)

Since x_B is an optimal solution of the L.P.P. (maximization problem)

$\therefore \Delta_j = c_j - Z_j \leq 0$, for all j not in the basis.

If Δc_k is the change in the cost c_k and c_k is not present in c_B (the vector of the costs associated to basic variables) then c_B remains unaltered with this change. Also there is no change in B.

$\therefore Z_j = c_B B^{-1} \alpha_j = c_B Y_j$ also remains unaltered.

If x_B is still an optimal solution of the given L.P.P. when c_k changes to $c_k + \Delta c_k$ then, $c_j - Z_j$ remains same for all j ($\neq$ k) not in the basis,

$\therefore$ We must have

$$(c_k + \Delta c_k) - Z_k \leq 0$$

or $\quad \Delta c_k \leq Z_k - c_k \; (= -\Delta_k).$ ...(1)

Hence if $c_k \notin c_B$ changes to $c_k + \Delta c_k$ such that $\Delta c_k \leq Z_k - c_k$ $(= -\Delta_k)$, the value of the objective function and the optimal solution of *the problem remains unchanged. Note that there is no lower bound to* Δc_k.

(ii) Variation in $c_j \in c_B$.

(i.e., Change in c_j Which is the Price of the Basic Variable)

We know that

$$Z_j = c_B B^{-1} a_j = c_B Y_j = \sum_{i=1}^{m} c_{Bi} Y_{ij}.$$

Let Δc_{Bk} be the change in c_{Bk}, the price of the basic variable x_{Bk} then if Z_j^* is the value of Z_j in this solution, then we have

$$Z_j^* = \sum_{\substack{i=1 \\ i \neq k}}^{m} c_{Bi} y_{ij} + (c_{Bk} + \Delta c_{Bk}) y_{kj}$$

$$= \sum_{i=1}^{m} c_{Bi} y_{ij} + y_{kj} \Delta c_{Bk} = Z_j + y_{kj} \Delta c_{Bk}$$

$$\therefore \quad c_j - Z_j^* = c_j - (Z_j + y_{kj} \Delta c_{Bk}) = (c_j - Z_j) - y_{kj} \Delta c_{Bk}.$$

The solution $\mathbf{x}_B$ will remain optimal for the change Δc_{Bk} in c_{Bk} if

$c_j - Z_j^* \leq 0$ for all j not in the basis

or $(c_j - Z_j) - y_{kj} \Delta c_{Bk} \leq 0$

or $y_{kj}, \Delta c_{Bk} \geq c_j - Z_j$

$$\therefore \quad \Delta c_{Bk} \geq \frac{c_j - Z_j}{y_{kj}}, \text{ for } y_{kj} > 0$$

$$\text{and } \Delta c_{Bk} \leq \frac{c_j - Z_j}{y_{kj}}, \text{ for } y_{kj} < 0. \quad ...(2)$$

Here the range of Δc_{Bk} *(change in the price of* c_{Bk} *corresponding tooth basis variable* x_{Bk}*) so that the solution remains optimal is given by*

$$\underset{y_{kj} > 0}{\text{Max}} \cdot \left[\frac{c_j - Z_j}{y_{kj}}\right] \leq \Delta c_{Bk} \leq \underset{y_{kj} < 0}{\text{Min}} \cdot \left[\frac{c_j - Z_j}{y_{kj}}\right] \quad ...(3)$$

for all j corresponding to which α_j is not in the optimal basis.

If no $y_{kj} > 0$*, there is no lower bound to* Δc_{Bk} *and if no* $y_{kj} < 0$*, there is no upper bound to* Δc_{Bk}.

To find the change in the value of the objective function.

The value of the objective function for the price vector c is given by

$$\mathbf{Z} = \mathbf{c_B} \cdot \mathbf{x_B} = \sum_{i=1}^{m} c_{Bi} \cdot x_{Bi}.$$

When $c_{Bk} \in c_B$ is changed to $c_{Bk} + \Delta c_{Bk}$, if Z^* is the value of the objective function then

$$Z^* = \sum_{\substack{i=1 \\ i \neq k}}^{m} c_{Bi} . x_{Bi} + (c_{Bk} + \Delta c_{Bk}) x_{Bk}$$

$$= \sum_{i=1}^{m} c_{Bi} . x_{Bi} + x_{Bk} . \Delta c_{Bk} = Z + x_{Bk} \, \Delta \, c_{Bk}.$$

Hence if Δc_{Bk} (change in c_{Bk}) satisfies (3), the solution $\mathbf{x}_B$ will remain optimal and the value of the objective function will be improved by an amount $x_{Bk} . \Delta c_{Bk}$ where x_{Bk} is the basic variable corresponding to c_{Bk}.

Example 1:

Find an optimal solution to the following L.P.P.

$$\text{Max. } Z = 15x_1 + 45x_2$$

$$\text{s.t. } x_2 \leq 50$$

$$x_1 + 1.6x_2 \leq 240$$

$$5x_1 + 2.9x_2 \leq 162$$

and x_1, x_2 ³ $0.$

If Max. $Z = \sum c_i x_i$ $(i = 1, 2)$ and c_2 is kept fixed at 45, find how much can c_1 be changed without affecting the above optimal solution.

Solution:

Introducing the slack variable x_3, x_4 and x_5, the given problem in standard form for simplex method is as follows:

$$\text{Max. } Z = 15x_1 + 45x_2 + 0.x_3 + 0.x_4 + 0.x_5$$

$$\text{s.t. } 0.x_1 + x_2 + x_3 = 50$$

$$x_1 + 1.6x_2 + x_4 = 240$$

$$5x_1 + 2.0x_2 + x_5 = 162$$

and $x_1, x_2, x_3, x_4, x_5 \geq 0.$

Taking $x_1 = 0 = x_2$, $x_3 = 50$, $x_4 = 240$, $x_5 = 162$, which is the starting B.F.S.

Solving the given L.P.P. by simple method, we get the following simplex table.

B	c_B	c_j x_B	15 Y_1	45 Y_2	0 Y_3	0 Y_4	0 Y_5	Min. ratio x_B/Y_2
Y_3	0	50	0	(1)	1	0	0	50/1
								←
Y_4	0	240	1	1.6	0	1	0	240/1.6
Y_5	0	162	.5	2	0	0	1	162/2
$Z = c_Bx_B = 0$		Δ_j	15	45	0	0	0	x_B/Y_1
				↑	↓			
Y_2	45	50	0	1	1	0	0	–
Y_4	0	160	1	0	–1.6	1	0	160/1
Y_5	0	62	(.5)	0	–2	0	1	26/.5
								←
Z = 2250	Δ_j	15	0	–45	0	1	x_B/Y_3	
		↑				↓		
Y_2	45	50	0	1	1	0	0	50/1
Y_4	0	36	0	0	(2.4)	1	–2	36/2.4
								←
Y_1	15	124	1	0	–4	0	2	–
Z = 4110	Δ_j	0	0	15	0	–30		
				↑	↓			
Y_2	45	35	0	1	0	–5/12	5/6	
Y_3	0	15	0	0	1	5/12	–5/6	
Y_1	15	184	1	0	0	5/3	–4/3	
Z = 4335	Δ_j	0	0	0	–25/4	–35/2		

From the table we set the optimal solution of the given L.P.P is

$x_1 = 184$, $x_2 = 35$, Max. $Z = 4335$.

To find variation in c_1. Here $c_1 \in c_B$.

From the above simplex table, we have

$$c_B = (c_{B1}, c_{B2}, c_{B3}) = (45, 0, 15) = (c_2, c_3, c_1)$$

$\therefore c_1 = c_{Bk} = c_{B3} = 15.$

$\therefore$ We have to find the range of variation in $c_1 \in c_B$.

Here we want to find the range Δc_{B3} (change in c_{B3}), which is given by (3), 2.

$$\underset{y_{kj}>0}{\text{Max}}\left[\frac{c_j - Z_j(=\Delta_j)}{y_{kj}}\right] \le \Delta c_{Bk} \le \underset{y_{kj}<0}{\text{Min}}\left[\frac{c_j - Z_j(=\Delta_j)}{y_{kj}}\right]$$

Here k = 3.

Here $y_{kj} = y_{3j} = (1, 0, 0, -4/3)$, out of which y_{31}, y_{32}, y_{33} are not to be taken j = 1, 2, 3 correspond to $\alpha_1, \alpha_2, \alpha_3$ which are in the basis. Thus here we shall consider y_{34} and y_{35} only.

Now $y_{34} = 5/3 > 0$ and $y_{35} = -4/3 < 0$.

$\therefore$ Δc_{B3} is given by

$$\text{Max.}\left[\frac{\Delta_4}{y_{34}}\right] \le \Delta c_{B3} \le \text{Min.}\left[\frac{\Delta_5}{y_{35}}\right]$$

or $$\text{Max.}\left[\frac{-25/4}{5/3}\right] \le \Delta c_{B3} \le \text{Min.}\left[\frac{-35/2}{-4/3}\right]$$

or $$-15/4 \le \Delta c_{B3} \le 105/8.$$

$\therefore$ Range of variation of $c_{B3} = c_1$ is given by

$$c_{B3} + (-15/4) \le c_1 \le c_{B3} + 105/8$$

or $$15 + (-15/4) \le c_1 \le 15 + 105/8$$

or $$45/4 \le c_1 \le 225/8.$$

If we take $c_1 = 45/4$, then proceeding as in above table, we do not get the same optimal solution.

$\therefore$ The variation in c_1 without affecting the above optimal solution is given by $45/4 < c_1 \le 225/8$.

Example 2:

The linear programming problem is

$$\text{Max } Z = 3x_1 + 5x_2$$

$$\text{s.t. } x_1 + x_2 \le 1$$

$$2x_2 + 3x_2 \le 1$$

$$\text{and } x_1, x_2 \ge 0.$$

Obtain the variations in cj (j = 1, 2) which are permitted without changing the optimal solution.

Solution:

Proceeding as usual, the *final simplex table* giving the optimal solution of the given L.P.P. is as follows.

		c_j	3	5	0	0
B	c_B	x_B	Y_1	Y_2 (β_2)	Y_3 (β_1)	Y_4
Y_3	0	2/3	1/3	0	1	–1/3
Y_2	5	1/3	2/3	1	0	1/3
$Z = c_B x_B = 5/3$		Δ_j	–1/3	0	0	–5/3

From the above table the optimal solution, we have $x_1 = 0$, $x_2 = 1/3$, Max. $Z = 5/3$.

Here $c_B = (c_{B1}, c_{B2}) = (0, 5) = (c_3, c_2)$.

To find variation in c_1. Here c_1 does not belong to c_B.

$\therefore$ From (1), 2, the change Δc_1 in c_1, so that the solution remains optimal, is given by

$$\Delta c_1 \le Z_1 - c_1 = -\Delta_1.$$

$$\Rightarrow \Delta c_1 \le 1/3.$$

$\therefore$ The range over which c_1 can vary maintaining the optimality of the solution given in above table is

$$-\infty < c_1 \le c_1 + \Delta c_1$$

$$\Rightarrow -\infty < c_1 \le 3 + 1/3$$

$$\Rightarrow -\infty < c_1 \le 10/3.$$

The value of the objective function will not change in this case.

To find variation in c_2. Here $c_2 = c_{Bk} = c_{B2} = 5 \in c_B$

$\therefore$ From (3), 2 the range of Δc_{B2} (change in c_{B2}) is given by

$$\underset{y_{kj}>0}{\text{Max}}\left[\frac{c_j - Z_j}{y_{kj}}\right] \le \Delta c_{Bk} \le \underset{y_{kj}<0}{\text{Min}}\left[\frac{c_j - Z_j}{y_{kj}}\right].$$

Here k = 2 and $y_{21} = 2/3$, $y_{24} = 1/3 > 0$. Here we cannot consider y_{22} and y_{23} as α_2 and α_3 are in the basis.

Since no $y_{2j} < 0$,

$\therefore$ There is no upper bound to Δc_{B2}.

The change in Δc_{B2} is given by

$$\text{Max.}\left[\frac{\Delta_1}{y_{21}}, \frac{\Delta_4}{y_{24}}\right] \le \Delta c_{b2} < \infty$$

$$\text{Max.}\left[\frac{-1/3}{2/3}, \frac{-5/3}{1/3}\right] \le \Delta c_{B2} < \infty$$

$\Rightarrow -1/2 \le \Delta c_{B2} < \infty$

$\Rightarrow c_{B2} - 1/2 \le c_2 < c_{B2} + \infty$

$\Rightarrow 5 - 1/2 \le c_2 < 5 + \infty$

$\Rightarrow 9/2 \le c_2 < \infty$.

Example 3:

Find an optimal solution to the following L.P. problem

$$Max.\ Z = 3x_1 + 5x_2$$

s.t. $x_1 \le 4$

$x_2 \le 6$

$3x_1 + 2x_2 \le 18$ *and* $x_1, x_2 \ge 0$,

what happens to this optimal solution if the objective is changed to

$$Z^* = 3x_1 + x_2.$$

Solution:

Introducing the slack variables x_3, x_4 and x_5 the given problem in standard form for simplex method is as follows:

$$\text{Max. } Z = 3x_1 + 5x_2 + 0.x_3 + 0.x_4 + 0.x_5$$

s.t $1.x_1 + x_3 = 4$

$0.x_1 + 1.x_2 + x_4 = 6$

$3x_1 + 2x_2 + x_5 = 18$

and $x_1, x_2, \ldots, x_5 \ge 0$.

Taking $x_1 = 0 = x_2$, we get $x_3 = 4$, $x_4 = 6$, $x_5 = 18$, which is the starting B.F.S.

Proceeding as usual we get the following simplex table.

B	c_B	c_j / x_B	3 Y_1	5 Y_2	0 Y_3	0 Y_4	0 Y_5	Min. ratio x_B/Y_2
Y_3	0	4	1	0	1	0	0	–
Y_4	0	6	0	(1)	0	1	0	6/1 ←
Y_5	0	18	3	2	0	0	1	18/2
Z = 0		Δ_j	3	5 ↑	0	0 ↓	0	x_B/Y_1
Y_3	0	4	1	0	1	0	0	4/1
Y_2	5	6	0	1	0	1	0	–
Y_5	0	6	(3)	0	0	–2	1	6/3 ←
Z = 30		Δ_j	3 ↑	0	0	–5	0 ↓	
Y_3	0	2	0	0	1	2/3	–1/3	
Y_2	5	6	0	1	0	1	0	
Y_1	3	2	1	0	0	–2/3	1/3	
Z = 36		Δ_j	0	0	0	–3	–1	

From the above table the

∴ Optimal solution of the given L.P.P. is

$x_1 = 2$, $x_2 = 6$ and Max. Z = 36.

Now the objective function is changed to $Z^* = 3x_1 + x_2$

i.e., c_2 is changed to 1 keeping c_1 fixed.

To find variation in c_2. Here $c_2 \in c_B$, and from last simplex table

$$c_B = (c_{B1}, c_{B2}, c_{B3}) = (0, 5, 3) = (c_3, c_2, c_1)$$

∴ $c_2 = c_{Bk} = c_{B2} = 5.$

∴ Range of Δc_{B2} is given by

$$\underset{y_{2j} > 0}{\text{Max}} \left[\frac{c_j - Z_j (= \Delta_j)}{y_{2j}} \right] \le \Delta c_{B2} \le \underset{y_{2j} < 0}{\text{Min}} \left[\frac{c_j - Z_j (= \Delta_j)}{y_{2j}} \right]$$

for all j not in the basis.

$$\Rightarrow \text{Max.} \left[\frac{\Delta_4}{y_{24}} \right] \leq \Delta c_{B2} < \infty$$

$$\Rightarrow -3 \leq \Delta c_{B2} < \infty$$

$$\therefore 5 - 3 \leq c_{B2} < 5 + \infty \text{ or } 2 \leq c_2 < \infty$$

$\therefore$ If c_2 is changed to 1, then the optimal solution given in the above table does not remain optimal.

To find the new optimal solution. When Z is changed to Z*.

If the objective function Z is changed to $Z^* = 3x_1 + x_2$ then c_B in the last simplex becomes (0, 1, 3).

$\therefore \quad \Delta_1 = 0 \; \Delta_2 = \Delta_3, \Delta_4 = c_4 - c_B Y_4 = 1, \Delta_5 = c_5 - c_B Y_5 = -1.$

Thus from the last of the above table, we get the new table as follows:

		c_j	3	1	0	0	0	Min. ratio x_B/Y_4
B	c_B	x_B	Y_1	Y_2	Y_3	Y_4	Y_5	
Y_3	0	2	0	0	1	(2/3)	–1/3	2/(2/3) ←
Y_2	1	6	0	1	0	1	0	6/1
Y_1	3	2	1	0	0	–2/3	1/3	
$Z^* = 12$	Δ_j		0	0	0	1 ↓	–1 ↑	
Y_4	0	3	0	0	3/2	1	–1/2	
Y_2	1	3	0	1	–3/2	0	1/2	
Y_1	3	4	1	0	1	0	0	
$Z^* = 15$	Δ_j		0	0	–3/2	0	–1/2	

$\therefore$ The solution of the new L.P.P. with changed objective function is $x_1 = 4$, $x_2 = 3$ and Max. $Z^* = 15$.

VARIATION IN THE REQUIREMENT VECTOR b

We know that the condition of optimality for the B.F.S. of a L.P.P. is $\Delta_j = c_j - Z_j \leq 0$.

Since Δ_j does not involve any of b_i if any component b_i of the requirement vector $b = [b_1, b_2, ..., b_m]$ is changed will not affect the conditions of optimality. Hence if any component b_i is changed to $b_i + \Delta b_i$ then the new solution thus obtained will remain optimal. But $x_B = B^{-1} b$ depends on b, therefore any change in b may affect the feasibility of the optimal solution *i.e.*, the optimal solution obtained by changing b may or may not be feasible.

Thus, a change in b_i must be of the magnitude vector b be changed to $b_j + \Delta b_j$, therefore if the new requirement vector is b*, then

$$b^* = [b_1, b_2, \ldots, b_l + \Delta b_l, \ldots, b_m].$$

If x_B^* is the solution of the new L.P.P. obtained by changing b_l to $b_t + Db_l$, then

$$x_B^* = B^{-1} b^* \text{ where B is the optimal basis.}$$

Let $\quad B^{-1} = (\beta_1, \beta_2, \ldots, \beta_m)$

$$= \begin{bmatrix} \beta_{11} & \beta_{12} & \cdots & \beta_{1t} & \cdots & \beta_{lm} \\ \beta_{21} & \beta_{22} & \cdots & \beta_{2t} & \cdots & \beta_{2m} \\ \vdots & \vdots & & \vdots & & \vdots \\ \beta_{i1} & \beta_{i2} & \cdots & \beta_{il} & \cdots & \beta_{im} \\ \vdots & \vdots & & \vdots & & \vdots \\ \beta_{ml} & \beta_{m2} & \cdots & \beta_{ml} & \cdots & \beta_{mm} \end{bmatrix}$$

l-th component

Since $b^* = [b_1, b_2, \ldots, b_l + \Delta b_l, \ldots, b_m]$

$$= [b_l + 0, b_2 + 0, \ldots, b_l + \Delta\, b_l, \ldots, b_m + 0]$$

$$= [b_1, b_2, \ldots, b_l, \ldots, b_m] + [0, 0, \ldots, \Delta\, b_l, \ldots, 0]$$

$$= b + [0, 0, \ldots, \Delta\, b_l, \ldots, 0]$$

$$\therefore \quad x_B^* = B^{-1}.\, b^*$$

$$= B^{-1}.\, \{b + [0, 0, \ldots, \Delta b_i, \ldots, 0]\}$$

$$= B^{-1} b + B^{-1}.\, [0, 0, \ldots, \Delta b_i, \ldots, 0]$$

l-th component

$$= x_B + [\beta_1, \beta_2, \ldots, \beta_l, \ldots, \beta_m].\, [0, 0, \ldots, \Delta b_l, \ldots, 0]$$

$$= x_B + \beta_l.\, \Delta b_l$$

$$= [x_{B1}, x_{B2}, \ldots, x_{Bl}, \ldots, x_{Bm}] + [\beta_{1l}, \beta_{2l}, \ldots, \beta_{ll}, \ldots, \beta_{ml}]\, .\, \Delta b_l.$$

$$= [x_{B1} + \beta_{11}\, \Delta b_l, \ldots, x_{Bl} + \beta_{ll}\, \Delta b_l, \ldots, x_{Bm} + \beta_{ml}.\, \Delta b_l] \quad \ldots(1)$$

Now if the solution x_B^* is feasible, then

$$x_{Bi} + \beta_{il}\, \Delta b_l \geq 0 \text{ for all } i = 1, 2, \ldots, m$$

$$\Rightarrow \quad \beta_{il}\, \Delta b_l \geq -\, x_{Bi}$$

$$\therefore \quad \Delta b_l \geq -\frac{x_{Bi}}{\beta_{it}}, \text{ for } \beta_{il} > 0$$

and $\quad \Delta b_l \leq -\dfrac{x_{Bi}}{\beta_{il}}$. for $\beta_{il} < 0$.

Hence the range of Δ_l so that the optimal solution x_B^* also remains feasible is given by

$$\underset{\beta_{il}>0}{\text{Max}}\left[\frac{x_{Bi}}{\beta_{il}}\right] \leq \Delta b_l \leq \underset{\beta_{il}<0}{\text{Min.}}\left[-\frac{x_{Bi}}{\beta_{il}}\right]$$

and the new value of the optimal solution is given by (1).

To find the change in the value of the objective function.

The given value of the objection function for a requirement vector is given by

$$Z = c_B\, x_B = \sum_{i=1}^{m} c_{Bi} \cdot x_{Bi}\,.$$

When b_l is changed to $b_l + \Delta b_l^*$ if Z^* is the value of the objective function, then $Z^* = c_B\, x_B$

$$= (c_{B1}, c_{B2}, \ldots, c_{Bm}).\ [x_{B1} + b_{1l}\, \Delta_{bl}, \ldots, x_{Bm} + b_{ml}\, \Delta b_l]$$

$$= \sum_{i=1}^{m} c_{Bi} \cdot (x_{Bi} + \beta_{il} \cdot \Delta b\,)$$

$$= \sum_{i=1}^{m} c_{Bi} \cdot x_{Bi} + \sum_{i=1}^{m} x_{Bi} \cdot \beta_{il} \cdot \Delta b_l$$

$$= Z + \sum_{i=1}^{m} x_{Bi} \cdot \beta_{il} \cdot \Delta b_l$$

Hence if Δb_l (change in b_l) satisfies (2), then the solution x_B^* given by (1) is also optimal feasible solution. and the value of the objective function is improved by an amount

$$\sum_{i=1}^{m} x_{Bi} \cdot \beta_{il} \cdot \Delta b_l\,.$$

Example:

Given the following L.P.P.

$$\text{Max. } Z = -x_1 + 2x_2 - x_3$$

s.t. $\quad 3x_1 + x_2 - x_3 \leq 10$

$\quad -x_1 + 4x_2 + x_3 \geq 6$

$\quad x_2 + x_3 \leq 4$

and $x_1, x_2, x_3 \geq 0.$

Find the separate range of b_1, b_2 and b_3 (the constants on the right hand sides of the constraints) consistent with the optimal solution.

Solution:

Introducing the slack variables x_4, x_6, surplus variables x_5 and artificial variables x_7, the given L.P.P. reduces to

$$\text{Max. } Z = -x_1 + 2x_2 - x_3 + 0.x_4 + 0.x_5 + 0.x_6 - Mx_7$$

$$\text{s.t. } 3x_1 + x_2 - x_3 + x_4 = 10$$

$$-x_1 + 4x_2 + x_3 - x_5 + x_7 = 6$$

$$x_2 + x_3 + x_6 = 4$$

and $x_1, \ldots, x_7 \geq 0.$

Proceeding as usual, the successive tables of simplex method are as follows

B	c_B	x_B	Y_1	Y_2	Y_3	Y_4	Y_5	Y_6	A_1	Min. ratio
		c_j	−1	2	−1	0	0	−M		x_B/Y_2
Y_4	0	10	3	1	−1	1	0	0	0	10/1
A_1	−M	6	−1	(4)	1	0	−1	0	1	6/4 ←
Y_6	0	4	0	1	1	0	0	1	0	4/1
$Z=c_B\ x_B=-6M$		Δ_j	−1 −M	2+4M ↑	−1+M	=	−M	0	0 ↓	x_B/Y_5
Y_4	0	17/2	13/4	0	−5/4	1	1/4	0		$\frac{17}{2}/\frac{1}{4}$
Y_2	2	3/2	−1/4	1	1/4	0	−1/4	0		$-\frac{5}{2}/\frac{1}{4}$ ←
Y_6	0	5/2	1/4	0	3/4	0	(1/4)	1		
$Z=c_B\ x_B=3$		Δ_j	−1/2	0	−3/2	0	1/2 ↑	0 ↓		
Y_4	0	6	3	0	−2	1	0	−1		
Y_2	2	4	0	1	1	0	0	−1		
Y_5	0	10	1	0	3	0	1	4		
$Z=c_B\ x_B=8$		Δ_j	−1	0	−3	0	0	−2		

We have Optimal solution is $x_1 = 0$, $x_2 = 4$ $x_3 = 0$, Max Z = 8.

Here $B = (\alpha_4, \alpha_2, \alpha_5)$, $b = (10, 6, 4)$

$x_B = (x_4, x_2, x_5) = (x_{B1}, x_{B2}, x_{B3}) = (6, 4, 10)$,

$$\therefore B^{-1} = (\beta_1, \beta_2, \beta_3) = \begin{bmatrix} 1 & 0 & -1 \\ 0 & 1 & 1 \\ 0 & 0 & 4 \end{bmatrix}$$

To find variation in b_1. After changing b_1 to $b_1 + \Delta b_1$, the new requirement vector is given by

$$b_B^* = (10 + \Delta b_1, 6, 4)$$

From relation (2), 3 the range of Δb_1 consistent with the optimal solution is given by

$$\underset{\beta_{i1} > 0}{\text{Max.}}\left[\frac{-x_{Bi}}{\beta_{i1}}\right] \le \Delta b_1 \le \underset{\beta_{i1} < 0}{\text{Min.}}\left[\frac{-x_{Bi}}{\beta_{i1}}\right]$$

Here $b_{11} = 1 > 0$. Since no $b_{i1} < 0$,

$\therefore$ there is no upper bound to Δb_1.

$\therefore$ We have Max. $\left[-\frac{x_{B1}}{\beta_{11}}\right] \le \Delta b_1 < \infty$.

$\Rightarrow -\frac{6}{1} \le \Delta b_1 < \infty$.

$\therefore$ Range for b_1 is $-6 + 10 \le b_1 < 10 + \infty$

$\Rightarrow 4 \le b_1 < \infty$

To find variation in b_2.

After changing b_2 to $b_2 + \Delta b_2$, the new requirement vector is $b_B^{**} = (10, 6 + \Delta b_2, 4)$.

From relation (2), 3 the range of Δb_2 consistent with the optimal solution is given by

$$\underset{\beta_{i2} > 0}{\text{Max}}\left[\frac{-x_{Bi}}{\beta_{i2}}\right] \le \Delta b_2 \le \underset{\beta_{i2} < 0}{\text{Min.}}\left[\frac{-x_{Bi}}{\beta_{i2}}\right]$$

Here $\beta_{22} = 1 > 0$. Since no $\beta_{i2} < 0$. $\therefore$ there is no upper bound to Δb_2.

$\therefore$ We have Max. $\left[-\frac{x_{B2}}{\beta_{22}}\right] \le \Delta b_2 < \infty$

$\Rightarrow \frac{-4}{1} \le \Delta b_2 < \infty$

$\therefore$ Range of variation of b_2 is

$$-4 + 6 \le b_2 < \infty + 0$$

$\Rightarrow$ $2 \le b_2 < \infty$.

To find variation in b_3.

After changing b_3 to $b_3 + \Delta b_3$, the new requirement vector is $b_B{}^{***} = (10, 6, 4 + \Delta_3)$, from relation (2), 3 the range of Δb_3 consistent with the optimal solution is given by

$$\underset{\beta_{i3}>0}{\text{Max.}}\left[\frac{-x_{Bi}}{\beta_{i3}}\right] \le \Delta b_3 \le \underset{\beta_{i3}<0}{\text{Min.}}\left[\frac{-x_{Bi}}{\beta_{i3}}\right]$$

Here $\beta_{23} = 1$, $\beta_{33} = 4 > 0$ and $\beta_{13} = -1 < 0$.

$$\therefore \quad \text{Max.}\left[\frac{-x_{B2}}{\beta_{23}}, -\frac{x_{B3}}{\beta_{33}}\right] \le \Delta b_3 \le \text{Min.}\left[-\frac{x_{B1}}{\beta_{13}}\right]$$

$$\Rightarrow \quad \text{Max.}\left\{-\frac{4}{1}, -\frac{10}{4}\right\} \le \Delta b_3 \le \text{Min.}\left\{\frac{-6}{-1}\right\}$$

$$\Rightarrow \quad -\frac{5}{2} \le \Delta b_3 \le 6$$

$\therefore$ Range of variation of b_3 is

$$4 - \frac{5}{2} \le b_3 \le 4 + 6$$

$$\Rightarrow \quad \frac{3}{2} \le b_3 \le 10.$$

VARIATION IN THE COMPONENT a_{ij} OF THE COEFFICIENT MATRIX A

Let the element a_{lk} of the l-th row and k-th column of the coefficient matrix A be changed to $a_{lk} + \Delta a_{lk}$. Now there are two possibilities according as a_{lk} is or not the element of the optimal basis B.

Case I. When a_{lk} is not an Element of Optimal Basis B

If a_{lk} is not an element of the optimal basis B, a change in a_{lk} will not affect B and so B^{-1} remains the same. Thus the change in such a_{lk} will not affect the solution $x_B = B^{-1}b$. Hence the feasibility of the solution x_B is not affected. But the change of a_{lk} may change the optimality of the solution *i.e.*, the solution x_B may not be the optimal solution of the new L.P.P. (obtained by changing a_{lk} to $a_{lk} + \Delta a_{lk}$). Thus we have to find the range of variation of a_{lk} so that the solution still remains optimal.

Since all the column voters except a_k of the matrix A remain unaffected by the variation Δa_{lk} in a_{lk}.

$\therefore \Delta_j = c_j - Z_j \leq 0$ for all j ($\neq$ k) not in the basis.

Thus the solution x_B will remain optimal for the new L.P.P. also, if

$$\Delta_k^* = c_k - Z_k^* \leq 0 \quad \text{...(1)}$$

where Δ_k^* and Z_k^* are Δ_k and Δ_k for the new L.P.P.

Let $\quad B^{-1} = (\beta_1, \beta_2, \ldots, \beta_l, \ldots, \beta_m)$

and $\quad \alpha_k = [a_{lk}, a_{2k}, \ldots, a_{lk}, \ldots, a_{mk}]$

If α_k^* is α_k for the new L.P.P. then

$$\alpha_k^* = [a_{lk}, a_{2k}, \ldots, a_{lk} + \Delta a_{lk}, \ldots, a_{mk}]$$
$$= [a_{1k} + 0, a_{2k} + 0, \ldots, a_{lk} + \Delta_{lk}, \ldots, a_{mk} + 0]$$
$$= [a_{1k}, a_{2k}, \ldots, a_{lk}, \ldots, a_{mk}] + [0, 0, \ldots, \Delta a_{lk}, \ldots, 0]$$
$$= \alpha_k + [0, 0, \ldots, \Delta a_{lk}, \ldots, 0]$$

l-th component

and $\quad Z_k^* = c_B \, B^{-1} \, \alpha_k^*$

$$= c_B . B^{-1} \{\alpha_k + [0, 0, \ldots, \Delta a^{lk}, \ldots 0]\}$$
$$= c_B . B^{-1} \alpha_k + c_B . B^{-1} [0, 0, \ldots, \Delta a_{lk}, \ldots, 0]$$
$$= Z_k + c_B . (\beta_1, \beta_2, \ldots, \beta_l, \ldots, \beta_m) . [0, 0, \ldots \Delta a_{lk}, \ldots, 0]$$
$$= Z_k + c_B (\beta_l \Delta a_{lk})$$
$$= Z_k + \Delta a_{lk} . c_B \beta_l$$
$$= Z_k + \Delta a_{lk} (c_{B1}, c_{B2}, \ldots, c_{Bl}, \ldots, c_{Bm}) \times [\beta_{1l}, \beta_{2l}, \ldots, \beta^{ll}, \ldots \beta_{ml}]$$
$$= Z_k + \Delta a^{lk} \sum_{i=1}^{m} c_{Bi} . \beta_{il} .$$

$\therefore$ From (1), the solution x_B will remain optimal if

$$\Delta^{k*} = c_k - Z_k^* = c_k - Z_k - \Delta a_{lk} \sum_{i=1}^{m} c_{Bi} . \beta_{il} \leq 0$$

$$\text{or } \Delta a^{lk} \sum_{i=1}^{m} c_{Bi} . \beta_{il} \; ^3 \; c^k - Z^k \; (= \Delta_k)$$

$$\therefore \Delta a_{lk} . \geq \frac{\Delta_k}{\sum_{i=1}^{m} c_{Bi} \beta_{il}} \text{ for } \sum_{i=1}^{m} c_{Bi} \beta_{il} > 0.$$

$$\text{and } \quad \Delta a_{lk} \leq \frac{\Delta_k}{\sum_{i=1}^{m} c_{Bi} \beta_{il}} \text{ for } \sum_{i=1}^{m} c_{Bi} \beta_{il} < 0.$$

Hence the range of Δa_{lk} (change in $a_{lk} \notin B$), so that the solution x_B remains optimal and feasible, is given by

$$\left[\frac{\Delta_k}{\left(\sum_{i=1}^{m} c_{Bi}\beta_{il}\right) > 0}\right] \le \Delta a_{lk} \le \left[\frac{\Delta_k}{\left(\sum_{i=1}^{m} c_{Bi}\beta_{il}\right) < 0}\right] \qquad ...(2)$$

If $\sum_{i=1}^{m} c_{Bi} \cdot \beta_{il} = 0$, Δa_{lk} is unrestricted.

If $\sum_{i=1}^{m} c_{Bi} \cdot \beta_{il} > 0$, there is no upper bound to Δa_{lk},

and if $\sum_{i=1}^{m} c_{Bi} \cdot \beta_{il} < 0$, there is no lower bound to Δa_{lk}.

Note: There is no change in the value of the objective function if D a_{lk} satisfies (2).

Case II. When a_{lk} is an Element of the Optimal Basis B

Since a_{lk} is an element of the optimal basis B then if a_{lk} is changed to $a_{lk} + \Delta a_{lk}$, the optimal basis B will certainly be changed and hence B^{-1} will also be changed. Consequently x_B = B–1 b and Zj = $c_B B^{-1} \alpha_j$ will also change. A change in Z_j may disturb the optimality condition $\Delta_j = c_j - Z_j \le 0$ while a change in x_B may disturb the feasibility of the solution. *Hence our aim is to find the range of variation of a_{lk} so that neither the optimality nor the feasibility of the solution is disturbed.* B

When the element $a_{lk} \in B$ is changed to $a_{lk} + \Delta a_{lk}$ then let the optimal basis B, the solution x_B and Z_j be represented by B^*, x_B^* and Z^*_j respectively.

Let $B = (b_1, b_2, \ldots, b_m)$ and α_k = bp

$\therefore \quad a_{lk} = b_{lp}$ and $a_{lk} + \Delta a_{lk} = b_{lp} + \Delta b_{lp}$

$$\therefore \quad B^* = B = \begin{pmatrix} 0 & 0 & \ldots & 0 & \ldots & 0 \\ 0 & 0 & \ldots & 0 & \ldots & 0 \\ \vdots & \vdots & & \vdots & & \vdots \\ 0 & 0 & \ldots & \Delta b_{lp} & \ldots & 0 \\ \vdots & \vdots & & \vdots & & \vdots \\ 0 & \ldots & 0 & \ldots & 0 & 0 \end{pmatrix} \leftarrow l-\text{th row}.$$

p-th column

$= B + \Delta b_{lp} \cdot 0_{lp}$, where 0_{lp} is the null matrix except for the (l, p)th element which equals to unity

$$= B\,(I + B^{-1}\,\Delta\,b_{lp}.\,0_{lp})$$

$$\therefore B^{*-1} = \{B(I + B^{-1}\,\Delta b_{lp}.\,0_{lp})\}-1 = (I + B^{-1}\,\Delta b_{lp}.\,0_{lp})^{-1}\,B^{-1} \qquad \text{...(3)}$$

$$[\because (AB)^{-1} = B^{-1}\,A^{-1}]$$

If $B^{-1} = (\beta_1, \beta_2, \ldots, \beta_m)$,

$$I + B^{-1}\,\Delta b_{lp}\,.0_{lp} = I + B^{-1}\begin{pmatrix} 0 & 0 & \cdots & 0 & \cdots & 0 \\ 0 & 0 & \cdots & 0 & \cdots & 0 \\ \vdots & \vdots & & \vdots & & \vdots \\ 0 & 0 & \cdots & \Delta b_{lp} & \cdots & 0 \\ \vdots & \vdots & & \vdots & & \vdots \\ 0 & \cdots & 0 & \cdots & 0 & 0 \end{pmatrix}$$

$$= I + \xrightarrow[\text{p-th row}]{} \begin{bmatrix} \beta_{11} & \beta_{12} & \cdots & \beta_{1l} & \cdots & \beta_{1m} \\ \beta_{21} & \beta_{22} & \cdots & \beta_{2l} & \cdots & \beta_{2m} \\ \vdots & \vdots & & \vdots & & \vdots \\ \beta_{pl} & \beta_{pl} & \cdots & \beta_{pl} & \cdots & \beta pm \\ \vdots & \vdots & & \vdots & & \vdots \\ \beta_{ml} & \beta_{m2} & \cdots & \beta_{ml} & \cdots & \beta_{mm} \end{bmatrix}$$

l-th column

$$\times \begin{pmatrix} 0 & 0 & \cdots & 0 & \cdots & 0 \\ 0 & 0 & \cdots & 0 & \cdots & 0 \\ \vdots & \vdots & & \vdots & & \vdots \\ 0 & 0 & \cdots & \Delta b_{lp} & \cdots & 0 \\ \vdots & \vdots & & \vdots & & \vdots \\ 0 & \cdots & 0 & \cdots & 0 & 0 \end{pmatrix} \xleftarrow[\text{l-th row}]{}$$

p-th column

$$= \xrightarrow[\text{p-th row}]{} \begin{pmatrix} 1 & 0 & \cdots & 0 & \cdots & 0 \\ 0 & 1 & \cdots & 0 & \cdots & 0 \\ \vdots & \vdots & & \vdots & & \vdots \\ 0 & 0 & \cdots & 1 & \cdots & 0 \\ \vdots & \vdots & & \vdots & & \vdots \\ 0 & 0 & \cdots & 0 & \cdots & 0 \end{pmatrix}$$

p-th column

$$+ \begin{pmatrix} 0 & 0 & \cdots & \beta_{1l}\Delta b_{lp} & \cdots & 0 \\ 0 & 0 & \cdots & \beta_{2l}\Delta b_{lp} & \cdots & 0 \\ \vdots & \vdots & & \vdots & & \vdots \\ 0 & 0 & \cdots & \beta_{pl}\Delta b_{lp} & \cdots & 0 \\ \vdots & \vdots & & \vdots & & \vdots \\ 0 & 0 & \cdots & \beta_{ml}\Delta b_{lp} & \cdots & 0 \end{pmatrix} \xleftarrow[\text{p-th row}]{}$$

p-th column

$$\Rightarrow x_B^* = \begin{bmatrix} x_{B1} \, 0 \dfrac{\beta_{1l} . \Delta b_{lp}}{D} . x_{Bp} \\ x_{B2} - \dfrac{\beta_{2l} . \Delta b_{lp}}{D} . x_{Bp} \\ \vdots \\ \dfrac{1}{D} x_{Bp} \\ x_{Bm} - \dfrac{\beta_{ml} . \Delta b_{lp}}{D} . x_{Bp} \end{bmatrix} \qquad ...(5)$$

The solution $\mathbf{x}_B^*$ is feasible if $\mathbf{x}_B^* \geq 0$.

$\therefore$ From (5), the solution x_B^* is feasible,

$$\text{if} \quad x_{Bi} - \frac{\beta_{il} . \Delta b_{lp}}{D} . x_{Bp} \leq 0 \qquad ...(6)$$

for all i = 1, 2, ... ,m $(i \neq p)$

$$\text{and } \frac{1}{D} . x_{Bp} \geq 0. \qquad ...(7)$$

Since $x_{Bp} \geq 0$

$\therefore$ (7) will hold of we have

$\therefore$ from (6), (assuming that (8) is true), we have

$$x_{Bi} \, D - \beta_{il} . \, \Delta b_{lp} \, x_{Bp} \geq 0$$

$$\Rightarrow x_{Bi} . \, (1 + \beta_{pl} \, \Delta b_{lp}) - \beta_{il} \, \Delta b_{pl} \, x_{Bp} \geq 0$$

$$\Rightarrow x_{Bi} \geq (\beta_{il} \, x_{Bp} - \beta_{pl} \, x_{Bi}) . \, \Delta b_{lp} \text{ for all } i \neq p$$

$$\therefore \Delta b_{lp} \leq \frac{x_{Bi}}{\beta_{il} x_{Bp} - \beta_{pl} x_{Bi}}, \text{ for } \beta_{il} \, x_{Bp} - \beta_{pl} \, x_{Bi} < 0$$

$$\text{and } \Delta b_{lp} \geq \frac{x_{Bi}}{\beta_{il} x_{Bp} - \beta_{pl} x_{Bi}}, \text{ for } \beta_{il} \, x_{Bp} - \beta_{pl} \, x_{Bi} < 0.$$

Hence the range of $\Delta a_{lk} (= \Delta b_{lp}) \in B$, so that the solution remains feasible, is given by

$$\text{Max.} \left[\frac{x_{Bi}}{P_i < 0} \right] \leq \Delta a_{lk} \leq \text{Min} \left[\frac{x_{Bi}}{P_i > 0} \right] \qquad ...(9)$$

where $D = 1 + \beta_{pl}\, \Delta\, \beta_{lp}, = 1 + \beta_{pl}\, \Delta\, a_{lk} > 0$. $Pi = \beta_{il}\, x_{Bp} - \beta_{pl}\, x_{Bi}$.

If no $P_i < 0$, there is no lower bound to $\Delta\, a_{lk}$ and if no $P_i > 0$, there is no upper bound to $\Delta\, a_{lk}$.

To Find the Range of Variation of a_{lk} for the Optimality of the Solution

For optimality of the solution $x_B{}^*$, we must have

$$\Delta_j{}^* = c_j - Z_j{}^* \leq 0$$

For all j not in the basis.

$$Z_j{}^* = c_B\, B^{*-1}\, \alpha_j = c_B\, (I + B^{-1}.\, \Delta\, b_{lp}\, 0_{lp})^{-1}.\, B^{-1}\, \alpha_j$$

$$= c_B\, (I + B^{-1}.\, \Delta\, b_{lp}\, 0_{lp})^{-1}.\, Y_j$$

Since $B^{-1}\, \alpha_j = Y_j = [y_{1j}, y_{2j}, \ldots, y_{pj}, \ldots, y_{mj}]$.

Using (4) and simplifying as before as in (1), we have

$$Z_j{}^* = c_B \begin{bmatrix} y_{1j} - \dfrac{\beta_{1l}\Delta b_{lp}}{D}.y_{pj} \\ y_{2j} - \dfrac{\beta_{2l}\Delta b_{lp}}{D}.y_{pj} \\ \vdots \\ \dfrac{1}{D} y_{pj} \\ \vdots \\ y_{mj} - \dfrac{\beta_{ml}\Delta b_{lp}}{D}.y_{pj} \end{bmatrix}$$

$$= (c_{B1}, c_{B2}, \ldots, c_{Bp}, \ldots, c_{Bm}). \begin{bmatrix} y_{1j} - \dfrac{\beta_{1l}\Delta b_{lp}}{D}.y_{pj} \\ y_{2j} - \dfrac{\beta_{2l}\Delta b_{pj}}{D} \\ \vdots \\ \dfrac{1}{D} y_{pj} \\ y_{mj} - \dfrac{\beta_{ml}\Delta b_{lp}}{D} y_{pj} \end{bmatrix} \begin{matrix} lp.y_{pj} \\ \longleftarrow \\ p\text{-th row} \end{matrix}$$

$$= \sum_{i=1}^{m} c_{Bi} \left(y_{ij} - \frac{\beta_{il}\Delta b_{lp}}{D} y_{pj} \right) + \frac{c_{By}.y_{pj}}{D}$$

$$= \sum_{i=1}^{m}\left(c_{Bi}y_{ij} - \frac{c_{Bi}\beta_{il}\Delta b_{lp}}{D}y_{pj}\right) - c_{Bp}\left(y_{pj} - \frac{\beta_{pl}\Delta b_{lp}}{D}y_{pj}\right) + \frac{c_{Bp}\cdot y_{pj}}{D}$$

$$= \sum_{i=1}^{m}c_{Bi}y_{ij} - \frac{1}{D}\sum_{i=1}^{m}c_{Bi}\beta_{il}\Delta b_{lp}y_{pj} - \frac{1}{D}[c_{Bp}\cdot(D - b_{pl}\,\Delta\,b_{lp})\,y_{pj} - c_{Bp}\,y_{pj}]$$

$$= Z_j - \frac{1}{D}\sum_{i=1}^{m}c_{Bi}\beta_{il}\Delta b_{lp}y_{pj}$$

Since $\sum_{i=1}^{m}c_{Bi}y_{ij} = Z_j$ and $D = 1 + \beta_{pl}\,\Delta b_{lp}$

$\therefore$ For optimality of the solution $x_B{}^*$

$$\Delta_j{}^* = c_j - Z_j{}^* = c_j - Z_j + \frac{1}{D}\sum_{i=1}^{m}c_{Bi}\cdot\beta_{il}\Delta b_{lp}y_{pj} \le 0 \;\forall\; j \text{ not in the basis}$$

$\Rightarrow$ $D\,(c_j - Z_j) + \sum_{i=1}^{m}c_{Bi}\beta_{il}\Delta b_{lp}y_{pj} \le 0$ assuming that $D = 1 + b_{pl}\,\Delta\,b_{lp} > 0$

$\Rightarrow$ $(1 + b_{pl}\,\Delta b_{lp})\,(c_j - Z_j) + \sum_{i=1}^{m}c_{Bi}\beta_{il}\Delta b_{lp}y_{pj} \le 0$

$\Rightarrow$ $\left[\beta_{pl}(c_j - Z_j) + \sum_{i=1}^{m}c_{Bi}\beta_{il}y_{pj}\right]\Delta\,b_{lp} \le -\,(c_j - Z_j)$

$\Rightarrow$ $Q_j\,\Delta\,b_{lp} \le -\,\Delta_j$

where $Q_j = b_{pl}\,(c_j - Z_j) + \sum_{i=1}^{m}c_{Bi}\beta_{il}y_{pj}$

$$= b_{pl}\,\Delta_j + \left[\sum_{i=1}^{m}c_{Bi}\beta_{il}\right]y_{pl}$$

$\therefore$ $\Delta\,b_{lp} \le \dfrac{-\Delta_j}{Q_j}$, for $Q_j > 0$

and $\Delta\,b_{lp} \ge \dfrac{-\Delta_j}{Q_j}$, for $Q_j < 0$.

Hence the range of $\Delta\,a_{lk}$ $(= \Delta\,b_{lp})$ *so that the solution remains optimal is given by*

$$\text{Max.}\left[-\frac{\Delta_j}{Q_j < 0}\right] \le \Delta\,a_{lk} \le \text{Min}\left[\frac{-\Delta_j}{Q_j > 0}\right] \qquad ...(10)$$

for all j not in the basis.

If no $Q_j < 0$, there is no lower bound to Δa_{lk}, and if no $Q_j > 0$, there is no upper bound to Δa_{lk}.

Hence a change Δa_{lk} in a_{lk} (an element of basis matrix B) for the solution to remain feasible and optimal can be made such that (8), (9) and (10) are satisfied.

Example:

Solve the following L.P.P.

$$\text{Max. } Z = 10x_1 + 3x_2 + 6x_3 + 5x_4$$

$$\text{s.t. } x_1 + 2x_2 + x_4 \leq 6$$

$$3x_1 + 2x_3 \leq 5$$

$$x_2 + 4x_3 + 5x_4 \leq 3$$

and $x_1, x_2, x_3, x_4 \geq 0.$

Compute the limits for a_{11} *and* a_{23} *so that the new solution remains optimal feasible solution.*

Solution:

The given L.P.P. in standard form can be written as follows:

$$\text{Max. } Z = 10x_1 + 3x_2 + 6x_3 + 5x_4 + 0.x_5 + 0.x_6 + 0.x_7$$

$$\text{s.t. } x_1 + 2x_2 + 0.x_3 + x_4 + x_5 = 6$$

$$3x_1 + 0.x_2 + 2.x_3 + 0.x_4 + x_6 = 5$$

$$0.x_1 + 1x_2 + 4.x_3 + 5.x_4 + x_7 = 3$$

and $x_1, x_2, \ldots, x_7 \geq 0.$

Proceeding as usual the final simplex table is as follows:

		c_j	10	3	6	5	0	0	0
B	c_B	x_B	Y_1	Y_2	Y_3	Y_4	Y_5	Y_6	Y_7
Y_2	3	56/27	0	1	−22/27	0	5/9	−5/27	−1/9
Y_1	10	5/3	1	0	2/3	0	0	1/3	0
Y_4	5	5/27	0	0	26/27	1	−1/9	1/27	2/9
$Z'=c_B\,x_B=643/27$		Δ_j	0	0	−82/27	0	−10/9	−80/27	−7/9

From the above table optimal solution of the given problem is given by

$$x_1 = 5/3,\ x_2 = 56/27,\ x_3 = 0,\ x_4 = 5/27 \text{ and } Z = 643/27.$$

$$\therefore\ B = (Y_2, Y_1, Y_4) = (y_1, y_2, y_3),\ c_B=(c_{B1}, c_{B2}, c_{B3})=(3, 10, 5)$$

$$x_B = (x_{B1}, x_{B2}, x_{B3}) = (56/27, 5/3, 5/27)\ \alpha_5\ \alpha_6\ \alpha_7$$

Since the initial (starting) basis matrix was $(Y_5\ Y_6\ Y_7)$:

from above table

$$B^{-1} = (\beta_1, \beta_2, \beta_3) = \begin{bmatrix} 5/9 & 5/27 & 1/9 \\ 0 & 1/3 & 0 \\ -1/9 & 1/27 & 2/9 \end{bmatrix}$$

i.e., $\beta_{11} = 5/9$, $\beta_{21} = 0$, $\beta_{31} = -1/9$,
$\beta_{12} = -5/27$, $\beta_{22} = 1/3$,
$\beta_{32} = 1/27$ and $\beta_{13} = -1/9$, $\beta_{23} = 0$, $\beta_{33} = 2/9$.

To Find Limits of Variation of a_{1k} for the Feasibility of the Solution

From (9) of 4, the range of variation of $a_{lk} = a_{11}$ for the feasibility of the solution is given by

$$\text{Max.}\left[\frac{x_{Bi}}{P_i < 0}\right] \le \Delta\, a_{lk} \le \text{Min.}\left[\frac{x_{Bi}}{P_i > 0}\right] \qquad \text{...(1)}$$

where $D = 1 + \beta_{pl}\,\Delta\,\beta_{lp} \ge 0$, $P_i = \beta_{il}\,x_{Bp} - \beta_{pl}\,x_{Bi}$.

Here $a_{lk} = a_{11} \in B$. Since $a_{11} \in \alpha_1\ (= Y_1) = y_2$

$\therefore\ a_{lk} = a_{11} = b_{12} = b_{lp}$ *i.e.*, $l = 1$ and $k = 1$, $p = 2$

and $i = 1, 2, 3$. (Since there are only three rows in A)

$x_{B1} = 56/27$, $x_{B2} = 5/3$, $x_{B3} = 5/27$.

$\therefore\ P_1 = \beta_{11}\,x_{B2} - \beta_{21}\,x_{B1} = (5/9).\,(5/3) - 0 = 25/27 > 0$

$P_2 = \beta_{21}\,x_{B2} - \beta_{21}\,x_{B2} = 0.5/3 - 0 = 0$

$P_3 = \beta_{31}\,x_{B2} - \beta_{21}\,x_{B3} = -\,1/9.\ 5/3 - 0 = -\,5/27 < 0.$

$\therefore$ from (1), we have

$$\text{Max.}\left[\frac{x_{B3}}{P_3 < 0}\right] \le \Delta\, a_{11} \le \text{Min.}\left[\frac{x_{B1}}{P_1 > 0}\right]$$

$$\Rightarrow \frac{5/27}{5/27} \le \Delta a_{11} \le \frac{56/27}{25/27}$$

$$\Rightarrow -1 \le \Delta\, a_{11} \le 56/25 \qquad \text{...(A)}$$

Also $D = 1 + \beta_{pl}\,\Delta\, a_{lk} = 1 \ge 0$, holds as $\beta_{pl} = \beta_{21} = 0$.

Again from (10) 4, the range of variation of $a_{lk} = a_{11}$ for the optimality of the solution is given by

$$\text{Max.}\left[\frac{-\Delta_j}{Q_j < 0}\right] \le \Delta a_{lk} \le \text{Min.}\left[\frac{-\Delta_j}{Q_j > 0}\right] \qquad \text{...(2)}$$

where $Q_j = \beta_{pl}\,\Delta_j + \left(\sum_{i=1}^{m} c_{Bi}\beta_{il}\right) y_{pj}$ for all j not in the basis

Here $Q_j = \beta_{21}\,\Delta_j + \left(\sum_{i=1}^{3} c_{Bi}\beta_{il}\right) y_{2j}$

$= \beta_{21}\,\Delta_j + (c_{B1}\,\beta_{11} + c_{B2}\,\beta_{21} + c_{B3}\,\beta_{31})\,y_{2j}$

$= 0 + (3.5/9 + 10.0 - 5.1/9)\,y_{2j} = (10/9)\,y_{2j}$ for all j = 3, 5, 6, 7 not in the basis.

$\therefore\quad Q_3 = 10/9.y_{23} = 10/9.2/3 = 20/27 > 0$

$Q_5 = 10/9.y_{25} = 10/9.0 = 0$

$Q_6 = 10/9.y_{26} = 10/9.1/3 = 10/27 > 0$

$Q_7 = 10/9.y_{27} = 10/9.0 = 0.$

Since no $Q_j < 0$

$\therefore$ There is no lower bound to $a_{lk} = (a_{11})$

$\therefore$ from (2), we have

$$-\infty < \Delta\,a_{11} \leq \text{Min.}\left\{-\frac{\Delta_3}{Q_3}, -\frac{\Delta_6}{Q_6}\right\}$$

$$\Rightarrow\ -\infty < \Delta\,a_{11} \leq \text{Min.}\left\{\frac{82/27}{20/27}, \frac{80/27}{10/27}\right\}$$

$\Rightarrow -\infty < \Delta\,a_{11} \leq 41/10.$ ...(B)

Since the range for Δa_{11} so that the solution remains optimal and feasible is given (A) and (B) both

$\therefore$ Both (A) and (B) are satisfied

if $-1 \leq \Delta\,a_{11} \leq 56/25.$

Since $a_{11} = 1$

$\therefore$ limits of variation of a_{11} are

$-1 + 1 \leq a_{11} \leq 56/25 + 1$

$\Rightarrow 0 \leq a_{11} \leq 81/25.$

To find limits of variation of a_{23}. Here $a_{23} \notin \alpha_3\ (Y_3)$ which is not in B. From (2), 4 the range of Δa_{lk} (change in $a_{lk} \notin$ B) so that the solution remains optimal and feasible is given by

$$\left[\frac{\Delta_k}{\left(\sum_{i=1}^{m} c_{Bi}\beta_{il}\right) > 0}\right] \leq \Delta a_{lk} \leq \left[\frac{\Delta_k}{\left(\sum_{i=1}^{m} c_{Bi}\beta_{il}\right)} < 0\right] \quad ...(3)$$

Here $a_{lk} = a_{23} = 2$

$\therefore\quad l = 2,\ k = 3.$

$\Delta_k = \Delta_3 = -82/27$, m = 3 (number of constraints)

$$\sum_{i=1}^{m} c_{Bi}\beta_{i1} = \sum_{i=1}^{3} c_{Bi}\beta_{i2} = c_{B1}\beta_{12} + c_{B2}\beta_{22} + c_{B3}\beta_{32}$$

$$= 3.(-5/27) + 10.(1/3) + 5(1/27) = 80/27 > 0$$

$\therefore$ There is no upper bound to Δa_{23}.

$\therefore$ from (3), we have

$$\frac{\Delta_3}{\sum_{i=1}^{3} c_{Bi}\beta_{i2} > 0} \leq \Delta a_{23} < \infty$$

$$\Rightarrow \frac{-82/27}{80/27} \leq \Delta a_{23} < \infty$$

$$\Rightarrow -41/40 \leq \Delta a_{23} < \infty.$$

Since $a_{23} = 2$,

$\therefore$ limits of variation of a_{23} are

$$-41/40 + 2 \leq a_{23} < \infty + 2$$

$$\Rightarrow 39/40 \leq a_{23} < \infty.$$

ADDITION OF A NEW VARIABLE TO THE PROBLEM

If a new variable is introduced in a L.P.P. whose optimal solution has been obtained, then the solution of the problem will remain feasible. Addition of an extra variable x_{n+1} to the problem will introduce an extra column say α_{n+1} to the coefficient matrix A and an extra cost c_{n+1} will be introduced in the price vector c. Thus the addition of this extra variable may affect the optimality of the problem.

For the same basis B the solution $\mathbf{x}_B$ of the original problem will remain optimal (maxima) if

$$c_{n+1} - Z_{n+1} \leq 0$$

$$\Rightarrow c_{n+1} - c_B B^{-1} \alpha_{n+1} \leq 0. \quad ...(1)$$

i.e., if (1) is satisfied then the new variable becomes just like a non-basis variable having zero value.

If $c_{n+1} - Z_{n+1} > 0$, then the solution $\mathbf{x}_B$ is no more optimal for the new problem and can be improved by introducing α_{n+1} in the basis. Here we can start with the last simplex table giving the optimal feasible solution of the original problem by introducing one more column corresponding to the variable x_{n+1}.

Example 1:

Solve the following L.P.P.

$$\text{Max. } Z = 3x_1 + 5x_2$$

$$\text{s.t. } x_1 + x_3 = 4$$

$$3x_1 + 2x_2 + x_4 = 18 .$$

and $x_1, x_2, x_3, x_4 \geq 0.$

If a new variable x_5 *is introduced in the above L.P.P. with price 7, then we have the following problem:*

$$\text{Max. } Z' = 3x_1 + 5x_2 + 7x_5$$

$$\text{s.t. } x_1 + x_3 + x_5 = 4$$

$$3x_1 + 2x_2 + x_4 + 2x_5 = 18$$

and $x_1, x_2, x_3, x_4, x_5 \geq 0.$

Find the solution of the new L.P.P.

Solution:

Proceeding as usual the successive simplex tables for the original L.P.P. are as follows:

Simplex Table 1

		c_j	3	5	0	0	Mini Ratio
B	c_B	x_B	Y_1	Y_2	Y_3	Y_4	x_B/Y_2
Y_3	0	4	1	0	1	0	–
Y_4	0	18	3	(2)	0	1	18/2 ←
$Z=c_B.x_B=0$		Δ_j	3	5 ↑	0	0 ↓	
Y_3	0	4	1	0	1	0	
Y_2	5	9	3/2	1	0	1/2	
$Z=c_B.x_B=45$		Δ_j	–9/2	0	0	–5/2	

∴ Optimal solution of the given L.P.P. is

$$x_1 = 0, x_2 = 9, x_3 = 4, x_4 = 0, \text{Max. } Z = 45.$$

From the final table the solution of the dual of the given L.P.P. is

$$w_1 = 0, w_2 = 5/2, \text{Min. } Z_D = 45.$$

Revised L.P.P. The dual of the new L.P.P. is same as that of the original L.P.P. with one more constraint $w_1 + 2w_2 \geq 7$ which corresponds to the new variable x_5.

The optimal solution of the dual of the original L.P.P. is $w_1 = 0$, $w_2 = 5/2$ which does not satisfy this constraint. Thus we see that the optimal solution of the dual of the given L.P.P. is not optimal for the revised problem and can be improved by introducing $\alpha_5 = [1, 2]$, (column of A corresponding to the new variable introduced) in the basis.

Thus, consider one more column Y_5 in the above table.

Since $B^{-1} = \begin{bmatrix} 1 & 0 \\ 0 & 1/2 \end{bmatrix}$

$$\therefore \quad Y_5 = B^{-1}\alpha_5 = \begin{bmatrix} 1 & 0 \\ 0 & 1/2 \end{bmatrix} \cdot \begin{bmatrix} 1 \\ 2 \end{bmatrix} = \begin{bmatrix} 1 \\ 1 \end{bmatrix}$$

$$\Delta_5 = c_5 - c_B Y_5 = 7 - (0, 5), (1, 1) = 2.$$

We get the following simplex table for revised L.P.P. as

Simplex Table 2

		c_j	3	5	0	0	7	Mini Ratio
B	c_B	x_B	Y_1	Y_2	Y_3	Y_4	Y_5	x_B/Y_5
Y_3	0	4	1	0	1	0	(1)	4/1 ←
Y_2	5	9	3/2	1	0	1/2	1	9/1
$Z'=c_B x_B=45$		Δ_j	–9/2	0	0 ↓	–5/2 ↑	2	
Y_5	7	4	1	0	1	0	1	
Y_2	5	5	1/2	1	–1	1/2	0	
$Z'=c_B x_B=53$		Δ_j	–13/2	0	–2	–5/2	0	

$\therefore$ Optimal solution of revised L.P.P. is

$x_1 = 0$, $x_2 = 5$, $x_3 = 0$, $x_4 = 0$, $x_5 = 4$.

and Max. $Z' = 53$.

ADDITION OF A NEW CONSTRAINT TO THE PROBLEM

Let a new constraint be introduced to a L.P.P. whose optimal solution has been obtained. Here we assume that the additional constraint does not introduce any new variable with non-zero-price.

Let Z^* be the optimal (maximal) value of the objective function for the new L.P.P. while it was Z for the original problem. Let $Z^* > Z$. Since the new optimal solution satisfies the first m constraints as well as the additional new constraint, therefore, it is also an optimal solution of the original problem which is contradiction to the fact that we already had an optimal solution to the original L.P.P. Here $Z^* \leq Z$. (Max.)

Thus, we have the following two cases:

Case I

If the optimal solution of the original L.P.P. satisfies the new constraint, it is also an optimal solution of the new L.P.P. In this case the adulterant is redundant.

Case II.

If the optimal solution of the original L.P.P. does not satisfy the new constraint, a new optimal solution of the new L.P. problem must be obtained as follows:

To Find the New Optimal Solution of the New Enlarged Problem

Let B and B_1 be the optimal basis of the original and the enlarged L.P. problem respectively. Clearly B_1 is the square matrix of order (m + 1) if B the square matrix of order m.

$\therefore$ We can write

$$B_1 = \begin{bmatrix} B & 0 \\ \alpha & \pm 1 \end{bmatrix} \quad ...(1)$$

The last column of B_1 corresponds to the slack, surplus or artificial vector associated with the additional new constraints and α is the row vector of the coefficients, in the new constraint, of the variables which correspond to the vectors in the optimal basis B.

Since B^{-1} exist and is known therefore the inverse of B_1 is given by

$$B_1^{-1} = \begin{bmatrix} B^{-1} & 0 \\ \mp \alpha B^{-1} & \pm 1 \end{bmatrix} \quad ...(2)$$

Let a_{m+1}, j be the coefficient of x_j in the new (m + 1)th constraint and α_j^* the column vector of the coefficients of x_j in the enlarged problem. Also if Y_j^* and Z_j^* are Y_j and Z_j for the new problem, then we have

$$Y_j^* = B_1^{-1}.\alpha_j^* = \begin{bmatrix} B^{-1} & 0 \\ \mp \alpha B^{-1} & \pm 1 \end{bmatrix} \begin{bmatrix} \alpha_j \\ a_{m+1.j} \end{bmatrix}$$

$$= \begin{bmatrix} Y_j \\ \mp \alpha B^{-1} \alpha_j \pm a_{m+1,j} \end{bmatrix}$$

$$\Rightarrow \quad Y_j^* = \begin{bmatrix} Y_j \\ \mp \alpha Y_j \pm a_{m+1,j} \end{bmatrix}$$

Now $Z_j^* = c_{B1}\ Y_j^* = (c_B, c_{B.m+1}) \begin{bmatrix} Y_j \\ \mp \alpha Y_j \pm a_{m+1,j} \end{bmatrix}$

or $\quad Z_j^* = c_{B1}\ Y_j + c_{B,\ m+1} \cdot (\pm\ \alpha Y_j \pm a_{m+1,j})$ ...(3)

(i) *If slack or surplus variable is introduced in the additional constraint.*

In this case $c_{B,m+1} = 0$

$\therefore$ from (3),

$$Z_j^* = c_B\ Y_j = Z_j$$

$\therefore c_j - Z_j^* = c_j - Z_j$

$\therefore c_j - Z_j^*$

remains unchanged is this case. Since the optimal solution of the original problem does not satisfy the new constraint, therefore the slack or surplus variable introduced in the new constraint is negative. *Hence we can apply the dual simplex algorithm to find an optimal feasible solution of the new problem.*

(ii) *If the artificial variable is introduced in the additional constraint i.e., if the additional constraint is a perfect equality.* In this case the additional vector is an artificial vector. Now there are two possibilities:

(a) *If the artificial variable in the basic solution is negative,* then assigning a price to the artificial variable we can use the dual simplex algorithm for the removal of the artificial variable from the basis.

(b) *If the artificial variable in the basis solution is positive,* then assigning a price –M to the artificial variable we can use the standard simplex method for the removal of the artificial variable from the basis. It is important to note that in this case $c_j - Z_j$ will be changed.

Note: If the addition of new downstream alters the nature of the problem, then the new problem must be solved as a fresh problem.

Example:

Consider the following table which presents an optimal solution to some linear programming problem.

Table

B	c_B	c_j	2	4	1	3	2	0	0	0
		x_B	Y_1	Y_2	Y_3	Y_4	Y_5	Y_6	Y_7	Y_8
Y_1	2	3	1	0	0	–1	0	0.5	0.2	–1
Y_2	4	1	0	1	0	2	1	–1	0	0.5
Y_3	1	7	0	0	1	–1	–2	5	–0.3	2
$Z = c_B x_B = 17$		Δ_j	0	0	0	–2	0	–2	–0.1	–2

If the additional constraint

$$2x_1 + 3x_2 - x_3 + 2x_4 - 4x_5 \leq 5$$

were annexed to the system, would there be any change in the optimal solution? Justify your answer.

Solution:

From the table the optimal solution of the given L.P. problem is

$$x_1 = 3,\ x_2 = 1,\ x_3 = 7,\ x_4 = 0 = x_5 = x_6 = x_7 = x_8$$

which also satisfies the new additional constraint

$$2x_1 + 3x_2 - x_3 + 2x_4 - 4x_5 \leq 5.$$

Thus the optimal solution of the given problem will not be changed if we introduce the above constraint to it.

Hence, the additional constraint is redundant and the optimal sol. of the given L.P.P. is also the optimal solution of the new L.P.P.

EXERCISE

1. Write a short note on sensitivity analysis.
2. Given that the problem: Max. Z = **cx** such that **Ax = b, x ≥ 0,** has an optimal solution; can one obtain a linear programming problem which has an unbounded solution changing b along?
3. The following table gives the optimal solution to a L.P.P.

   Max. $Z = 3x_1 + 5x_2 + 4x_3,$

   s.t. $2x_1 + 3x_2 \leq 8,\ 2x_2 + 5x_3 \leq 10,$

   $3x_1 + 2x_2 + 4x_3 \leq 15.\ x_1, x_2, x_3 \geq 0.$

   For the above L.P.P. calculate the following:

(i) How much c_3 and c_4 can be increased before the present basic solution will no longer be optimal? Also, find the change in the value of the objective function if possible.

(ii) How much b_2 can be changed maintaining the feasibility of the solution?

(iii) Find the limits for the changes in a_{14} and a_{24} so that the new solution remains optimal feasible solution.

Table

B	c_B	x_B	Y_1	Y_2	Y_3	Y_4	Y_5	Y_6
Y_2	5	50/41	0	1	0	15/41	8/41	–10/41
Y_3	4	62/41	0	0	1	–6/41	5/41	4/41
Y_1	3	89/41	1	0	0	–2/41	–12/41	15/41
$Z = c_B x_B = 765/41$		Δ_j	0	0	0	–45/41	–24/41	–11/41

4. For the L.P.P. Max. $Z = 5x_1 + 3x_2$

Subject to, $3x_1 + 5x_2 \leq 15$

$5x_1 + 2x_2 \leq 10$

and $x_1, x_2 \geq 0$.

Find an optimal solution. Hence find how far the component c_1 of the vector c of the function $\mathbf{Z} = \mathbf{cx}$ can be increased without destroying the optimality of the solution.

5. Find an optimal solution to the following L.P.P.

Max. $Z = 15x_1 + 45x_2$

s.t. $x_1 + 16x_2 \leq 240$

$5x_1 + 2x_2 \leq 162$

$x_2 \leq 50,\ x_1, x_2 \geq 0.$

If max. $Z = \Sigma c_j x_j$, $j = 1, 2$ and c_2 is fixed at 45, determine how much c_1 can be changed without affecting the above solution.

6. Consider the following table which presents an optimal solution to some L.P.P.

For the above problem assuming that Y_4 and Y_5 were, in that order, in the initial identify matrix basis, calculate the following:

(i) How much can b_1 and b_2 be increase without affecting the optimality and feasibility of the solution?

(ii) How much c_3 can be increased before the present basic solution will no longer be optimal?

B	c_B	x_B	Y_1	Y_2	Y_3	Y_4	y_5
Y_1	2	1	1	0	0.5	4	–0.5
Y_2	3	2	0	1	1	–1	2
		x_j	1	2	0	0	0
Z = 8		c_j	2	3	1	0	0
		Δ_j	0	0	–3	–5	–5

s.t. $x_1 + 3x_2 - x_3 + 2x_5 - 7$

$-2x_2 + 4x_3 + x_4 = 12$

$-4x_2 + 3x_3 + 8x_5 + x_6 = 10$

and $x_j \geq 0$, $j = 1, 2, \ldots, 6$.

For the above L.P.P. calculate the following:

(i) Limits of variations of the costs c_1, c_2, c_3, c_4, c_5 and c_6, for which the optimal solution remains optimal.

(ii) Limits of variation of b_2 without affecting the optimality of the solution.

8. Solve the following L.P.P.

Max. $Z = 3x_1 + 5x_2$

s.t. $x_1 + x_3 = 4$

$3x_1 + 2x_2 + x_4 = 18$

and $x_1, x_2, x_3, x_4 \geq 0$.

Would there be any change in the optimal solution if the additional constraint (1) $x_2 \leq 10$ or (ii) $x_2 \leq 6$ is added to the above L.P.P. In case the optimal solution change find the new optimal feasible solution.

9. (i) Discuss the effect of discrete changes in the requirements (on the right side of the inequalities) for the following L.P.P.

Max. $Z = 3x_1 + 4x_2 + x_3 + 7x_4$

subject to

$8x_1 + 3x_2 + 4x_3 + x_4 \leq 7$

$2x_1 + 6x_2 + x_3 + 5x_4 \leq 3$

$x_1 + 4x_2 + 5x_3 + 2x_4 \leq 8,$

$x_1\ x_2, x_3, x_4 \geq 0.$

(ii) Discuss the effect of discrete changes in c on the optimality of an optimum basis feasible solution to the above L.P.P.

10. Discuss the effect of adding a new non-negative variable x_8 in the L.P.P. of question 9 (i), on the optimality of its optimum. It is given that the coefficient of x_8 in the constraints of the problem are 2, 7 and 3 respectively, the cost component associated with x_8 being 5.

(ii) Explain the situation when we have $c_8 = 10$ instead of 5.

11. Consider the L.P.P.

$$\text{Mini } Z = x_2 - 3x_3 + 2x_5$$

$$\text{s.t.} \quad 3x_2 - x_3 + 2x_5 \le 7$$

$$-2x_2 + 4x_3 \le 12$$

$$-4x_2 - 3x_3 + 8x_5 \le 10,$$

$$x_2, x_3, x_5 \ge 0.$$

The optimal table is given as follows.

B	c_B	x_B	Y_1	Y_2	Y_3	Y_4	Y_5	Y_6
Y_2	–1	4	2/5	1	0	1/10	4/5	0
Y_3	3	5	1/5	0	1	3/10	2/5	0
Y_6	0	11	1	0	0	–1/2	10	1
		x_j	0	4	5	0	0	11
$Z'=c_B\, x_B=11$		c_j	0	1	–3	0	2	0
$Z=-11$		Δ_j	–1/5	0	0	–4/5	–12/5	0

(a) Formulate the dual problem for this primal problem.

(b) What are the optimal values of dual variables?

(c) How much c_5 be decreased before Y_5 goes in to basis?

(d) How much can the 7 in first constraint be increased before the basis would change?

12. Solve the following L.P.P. and find how much can c_1 be changed without affecting the optimal conditions.

$$\text{Max. } Z = 15x_1 + 45x_2 = c_1 x_1 + c_2 x_2$$

$$\text{s.t.} \quad x_1 + 1.6x_2 \le 240$$

$$.5x_1 + 2x_2 + x_3 = 162,$$

$$x_2 + x_4 = 50,$$

$$x_1, x_2, x_3, x_4 \ge 0.$$

13. Consider the following table which presents optimal solution to some linear programming problem. If the additional constraint $2x_1 + 3x_2 - x_3 + 2x_4 + 4x_5 \le 5$ were annexed to the system, would there be any change in the optimal solution?

B	c_B	x_B	Y_1	Y_2	Y_3	Y_4	Y_5	Y_6	Y_7	Y_8
Y_1	2	3	1	0	0	−1	0	0.5	0.2	−1
Y_2	4	1	0	1	0	2	1	−1	−1	0.5
Y_3	1	7	0	0	1	−1	−1	5	0.3	2
Z = 17		Δ_j	0	0	0	2	0	2	0.1	2

14. Consider the following optimal simplex table for a maximization problem (with all constraints of ≤ type), where x_4 is slack and a_1 is an artificial variable. Let a new variable $x_5 \ge 0$ be introduced in the problem with a cost 30 assigned to it in the objective function.

Also suppose that the coefficients of x_5 in the two constraints are 5 and 7 respectively.

B	c_B	x_B	Y_1	Y_2	Y_3	Y_4	A_1
Y_2	12	8/5	0	1	−1/5	2/5	−1/5
Y_1	5	9/5	1	0	7/5	1/5	2/5
Z = 141/5		Δ_j	−3	0	−3/5	−29/5	−M + 2/5

Discuss the effect of this new variable on the optimality of the given problem.

ANSWERS

3. (i) $5/4 \le c_3 \le 23/2$, value of objective function $= \frac{765}{41} + (\Delta c_3)\cdot\frac{62}{41}$ – $\infty < D\, c_4 \le 45/41$, no change in the value of obj. function

(ii) $-25/4 \le \Delta b_2 \le 89/12$.

(iii) $-1 \le \Delta a_{14} < \infty$, $-1 \le \Delta a_{24} < \infty$.

4. $x_1 = 20/19$, $x_2 = 45/19$, Max. Z = 235/19, $9/5 \le c_1 \le 15/2$

5. $x_1 = 352/13$, $x_2 = 173/13$, Max. Z = 1005, $45/16 < c_1 \le 225/2$

6. (i) $-1/4 \le \Delta b_1 \le 2$, $-1 \le \Delta b_2 \le 2$,

(ii) $c_3 \le 4$.

7. $x_1 = 0$, $x_2 = 4$, $x_3 = 5$, $x_4 = 0$, $x_5 = 0$, $x_6 = 11$, Max. Z = 11.

(i) $c_1 \leq 1/5$, $-3/2 \leq c_2 < \infty$, $2 \leq c_3 < \infty$, $c_4 \leq 4/5$, $c_5 \leq 2/5$, $-1/5 \leq c_6 \leq 8/5$.

(ii) $-14/3 \leq b_2 \leq 34$.

8. $x_1 = 0$, $x_2 = 9$, $x_3 = 4$, $x_4 = 0$, Max. Z = 45

(i) No change

(ii) $x_1 = 2$, $x_2 = 6$, $x_3 = 2$, $x_4 = 0$ Max. Z = 36.

9. (i) $-32/5 \leq \Delta b_1 \leq 5$, $-5/4 \leq \Delta b_2 \leq 84/5$, $-126/19 \leq \Delta b_3$, having no upper bound.

(ii) $-1/5 \leq \Delta c_1 \leq 63$, $\Delta c_2 \leq 169/38$,

$\Delta c_3 \leq 1/2$, $-169/42 \leq \Delta c_4 \leq 1/2$,

$\Delta c_5 \leq 1/38$, $\Delta c_6 \leq 53/38$, $-1/9 \leq \Delta c_7 \leq 1$.

11. (a) Mini $Z = 7w_1 + 12w_2 + 10w_3$,

s.t. $-3w_1 + 2w_2 = 4w_3 \leq 1$, $-w_1 + 4w_2 + 3w_3 \geq 3$,

$-2w_1 - 8w_3 \leq 2$, $w_1, w_2, w_3 \geq 0$

(b) $w_1 = 1/5$, $w_2 = 4/5$, $w_3 = 0$

(c) change c_5 to $c_5 - \Delta c_5$, s.t. $\Delta c_5 > 12/5$

(d) 7 changes to $7 + \Delta b_1$, s.t. $-10 \leq \Delta b_1 < \infty$.

12. $x_1 = 184$, $x_2 = 35$, $x_3 = 0$, $x_4 = 15$, Max. Z = 4335
$45/4 < c_1 \leq 225/8$.

13. $x_B^* = (3173)$.

14. $x_1 = 0$, $x_2 = 25/19$, $x_3 = 0$, $x_4 = 9/19$, Max. Z = 30.

EXERCISE
(Objective Questions)

Fill in the Blanks:

Fill in the blanks '...', so that the following statements are complete and correct.

1. The range of Δc_{Bk}, $c_{Bk} \in \Delta c_B$, so that the solution $x_B = B^{-1} b$ remains optimal is ...

2. If $c_{Bk} \in c_B$ lies in the range

$$\underset{y_{kj} > 0}{\text{Max.}} \left[\frac{c_j - Z_j}{y_{kj}} \right] \leq \Delta c_{Bk} \leq \underset{y_{kj} > 0}{\text{Min.}} \left[\frac{c_j - Z_j}{y_{kj}} \right],$$

the value of the objective function is improved by an amount...

3. The optimal solution of the L.P.P. remains ... when $c_k \notin c_B$ changes to $c_k + \Delta c_k$

4. If Δb_l lies in the range

$$\underset{\beta_{il}>0}{\text{Max.}}\left[-\frac{x_{Bi}}{\beta_{il}}\right] \le \Delta b_l \le \underset{\beta_{il}>0}{\text{Min.}}\left[-\frac{x_{Bi}}{\beta_{il}}\right],$$

the value of the objective function is improved by an amount...

5. The rage of Δb_l, so that the new optimal solution also remains feasible is ...

6. There is no change in the value of the objective function if Δa_{lk}, $a_{lk} \notin B$ satisfies ...

7. The range of Δa_{lk}, $a_{lk} \notin B$ so that the solution x_B remains optimal and feasible for the new L.P.P. is ...

8. The optimal solution obtained by changing the requirement vector b will not be feasible.

9. If $a_{lk} \notin B$, a change in a_{lk} will not affect the solution $x_B = B^{-1}$ b and it will be a solution of the new L.P.P.

10. Any change in $b_i \in b$ will change the condition of optimality.

11. If $a_{lk} \in B$, any change in a_{lk} will not change the solution $x = B^{-1}$ b as well as the objective function.

12. The value of the objective function does not change when $c_k \notin c_B$ changes to $c_k + \Delta c_k$.

13. The basis feasible solution $\mathbf{x}_B = B^{-1}\,\mathbf{b}$ of the L.P.P.

    Max. Z = **cx** s.t. **Ax** = **b**, **x** ≥ **0** will not change if we change some $c_j \in c$.

ANSWERS

1. $$\underset{y_{kj}>0}{\text{Max.}}\left[\frac{c_j - Z_j}{y_{kj}}\right] \le \Delta c_{Bk} \le \underset{y_{kj}>0}{\text{Min.}}\left[\frac{c_j - Z_j}{y_{kj}}\right]$$

2. $x_{Bk}\ \Delta_k$.

3. Unchanged.

4. $\sum_{i=1}^{m} c_{Bi}\beta_{il}\Delta b_l$.

SENSITIVITY ANALYSIS IN ASSIGNMENT PROBLEMS

There is very little scope for sensitivity analysis in assignment problem because of its structure. Modest alterations in the conditions (such as one being able to do two jobs) can be considered by repeating the man's row and adding a dummy a column to square up the matrix. The problem of not assigning a particular job (ith) to particular facility (jth) can be solved by taking a very large cost (∞) of this assignment.

Also, the addition of a constant throughout any row or column does not affect the optimal solution of the assignment problem.

THE TRAVELLING SALESMAN (ROUTING) PROBLEM

Suppose a salesman wants to visit a certain number of cities.

He knows the distances (or time or cost) of journey between every pair of cities allotted to him. His problem is to choose such a rout which starts from his home city, passes through each city once and only once and returns this home city in the shortest possible distance (or in least time or at least cost).

The above problem may be classified in two forms:

(i) Symmetrical

If the distance (or time or cost) between every pair of cities is independent of the direction of journey the problem is said to be symmetrical.

(ii) Asymmetrical

If for one or more pairs of cities, the distance (or time or cost) changes with the direction, the problem is said to be asymmetrical. For example, it takes longer time while going up-hill from city A to B instead of coming down hill from city B to A. Similarly flying from East to West usually takes longer time than from West to East on account of prevailing winds.

Further we note that for two cities there is only one possible route *i.e.*, there is no choice.

In case of three cities, say A, B and C, one of them (say A) is the home base, there are two possible routes: A → B → C and A → C → B. For four cities there are 3! = 6 possible routes. In general, there are $(n - 1)!$ possible routes if there are n cities. Thus, practically it is impossible to find the best route by trying each one. That is why the travelling salesman problem is considered as a puzzle by the mathematicians. The best procedure to solve the problem is as if it were an assignment problem. We formulate the problem of travelling salesman in the form of an assignment problem with the additional restriction on his choice of route.

Formulation of a Travelling Salesman Problem as Assignment Problem

Let $x_{ij} = 1$, if the salesman goes directly from city A_j, to c_{ij} to A_j and zero otherwise. Also, let c_{ij} be the distance (or time or cost) from city A_i to city A_j. Then our problem is to minimize $Z = \sum\sum c_{ij}\, x_{ij}$ with one additional restriction that the x_{ij}'s must be so chosen that no city is visited i_j twice until the tour of all the cities is completed. In particular, he cannot go directly from city A_i to A_i itself. To avoid this possibility in the minimization process we adopt the convention $c_{ij} = \infty$ so that x_{ij} can never be unity when $i = j$. Also we note that only one $x_{ij} = 1$ for each value of i and j. The distance (or time or cost) matrix for this problem is given in the following table:

		To			
		A_1	A_2	...	A_n
	A_1	∞	c_{12}	...	c_{1n}
	A_2	c_{21}	∞	...	c_{2n}
	.	.	.	.	.
From	.	.	.	.	.
	.	.	.	.	.
	A_n	c_{n1}	c_{n2}	...	∞

We can omit the variable x_{ij} from the problem specification. Our problem is to determine a set of n elements of this matrix, one in each row and one in each column, so as to minimize the sum of the elements determined above.

Note: A problem similar to travelling salesman arises when n items say A_i, i = 1, 2, ... n, are to be produced on a machine in continuation, given that c_{ij} (i, j = 1, 2, ... n) is the setup cost of the machine when item A_i is followed by A_j. Here two additional restrictions are imposed. One restriction is that we do not follow A_i again by A_i. The other restriction is that we do not produce an item again until all items are produced once.

Solution Procedure

The problem could be solved by assignment technique. In some cases (violation of additional restriction) we use the method of numeration by assignment the next minimum element of matrix in place of zero.

The following example will make the procedure clear.

Example 1:

Solve the travelling salesman problem given by the following data:

$$c_{12} = 20,\ c_{13} = 4,\ c_{14} = 10,\ c_{23} = 5,\ c_{34} = 6$$

$$c_{25} = 10,\ c_{35} = 6,\ c_{45} = 20,\ \text{where } c_{ij} = c_{ji},$$

and there is no route between cities i and j if the value for c_{ij} is not given above.

Solution:

Consider the given problem as an assignment problem.

Taking $c_{ij} = \infty$ for i = j, the cost matrix is as follows:

	1	2	3	4	5
1	∞	20	4	10	∞
2	20	∞	5	∞	10
3	4	5	∞	6	6
4	10	∞	6	∞	20
5	∞	10	6	20	∞

If there is no route between cities i and j then we have taken $c_{ij} = \infty$ to avoid the possibility of going from ith station to jth station.

Now we shall solve the problem by usual assignment algorithm. The following tables show the necessary steps for reaching the solution:

∞	15	(0)	4	∞
15	∞	⊗	∞	3
(0)	⊗	∞	⊗	⊗
4	∞	⊗	∞	12
∞	3	⊗	12	∞

∞	12	(0)	1	∞
12	∞	⊗	∞	(0)
(0)	⊗	∞	⊗	⊗
1	∞	⊗	∞	9
∞	(0)	⊗	9	∞

∞	12	(0)	⊗	∞
11	∞	⊗	∞	(0)
⊗	1	∞	(0)	1
(0)	∞	⊗	∞	9
∞	(0)	⊗	8	∞

Hence, the optimum solution of the assignment problem is:

$$1 \rightarrow 3,\ 3 \rightarrow 4,\ 4 \rightarrow 1,\ 2 \rightarrow 5,\ 5 \rightarrow 2.$$

But this is not the solution to the travelling salesman problem, as it is not allowed to go from city 4 to 1 without visiting the cities 2 and 5.

We try to find the 'next best' solution which satisfies the additional restriction. The smallest element other than zero is 1. So we try to bring 1 into the solution. Since the element 1 occurs at two places, we shall consider both the cases separately until the acceptable solution is attained.

Make assignment in the cell (3, 2) having the element 1 instead of making zero assignment in the cell (5, 2). After making this assignment we observe that no other assignment can be made in the third row and second column. Consequently make assignment in the cell (5, 4) having element 8, instead of zero assignment. Consequently make assignment in the cell (5, 2).

The new assignment plan is shown in the following table:

To

From		1	2	3	4	5
	1	∞	12	(0)	⊗	∞
	2	11	∞	⊗	∞	(0)
From	3	⊗	(1)	∞	⊗	1
	4	(0)	∞	⊗	∞	9
	5	∞	⊗	∞	(8)	∞

Thus the resulting feasible solution is

$$1 \to 3 \to 2 \to 5 \to 4 \to 1 \text{ with cost } 9.$$

Again if we make assignment in the cell (3, 5) having the next best element 1 instead of 0 marked in cell (3, 4), then no feasible solution is obtained in terms of zeros or with cost less than 9.

Hence the optimum route is $1 \to 3 \to 2 \to 5 \to 4 \to 1$

The corresponding cost = 4 + 5 + 10 + 20 + 10 = 49.

Example 2:

Given the matrix of setup costs, show how to sequence the production so as to minimize the setup cost per cycle.

To

From		A_1	A_2	A_3	A_4	A_5
	A_1	∞	2	5	7	1
	A_2	6	∞	3	8	2
From	A_3	8	7	∞	4	7
	A_4	12	4	6	∞	5
	A_5	1	3	2	8	∞

Solution:

Consider the problem as an assignment. Applying the assignment technique, we get the following matrix, showing a solution in terms of marked '□' zero:

To

From		A_1	A_2	A_3	A_4	A_5
	A_1	∞	1	3	6	(0)
	A_2	4	∞	(0)	6	⊗
From	A_3	4	3	∞	(0)	3
	A_4	8	(0)	1	∞	1
	A_5	(0)	2	⊗	7	∞

The solution to the assignment problem given by above matrix is

$A_1 \to A_2$, $A_5 \to A_1$, $A_2 \to A_3$, $A_3 \to A_4$, $A_4 \to A_2$.

This solution indicates to produce the products A_1, then A_5 and then again A_1, without producing the products A_2, A_3 and A_4, which violates the additional restriction of prodding each product once and only once before returning to the first product. So this is not a solution of the travelling salesman problem.

Now we try to find the next best solution which also satisfies the additional restriction. The next minimum element (non-zero) in the matrix is 1. We try to bring 1 in the solution. The cost 1 also occurs at three places.

Start by making unity-assignment in the cell (1, 2) instead of zero assignment in the cell (1, 5). Then no other assignment can be made in the first row and the second column. The best solution of the problem lies in the marked '□' elements as shown in the table.

		A_1	A_2	A_3	A_4	A_5
	A_1	∞	(1)	3	6	⊗
	A_2	4	∞	(0)	6	⊗
From	A_3	4	3	∞	(0)	3
	A_4	8	⊗	1	∞	(1)
	A_5	(0)	2	⊗	7	∞

Thus, the required solution of the problem is

$A_1 \to A_2 \to A_3 \to A_4 \to A_5 \to A_1$.

For this solution, the cost in the reduced matrix is 2.

On the other hand if we select the element 1 in the cell (4, 3) in the solution, then no feasible solution is available in terms of zero or for which the reduced matrix gives the minimum cost less than 2.

Hence, the most suitable sequence is $A_1 \rightarrow A_2 \rightarrow A_3 \rightarrow A_4 \rightarrow A_5 \rightarrow A_1$.

The minimum setup cost = 2 + 3 + 4 + 5 + 1 = 15.

Example 3:

A salesman estimate that the following would be the cost on his route visiting the six cities as shown.

		To city					
		1	*2*	*3*	*4*	*5*	*6*
	1	∞	*20*	*23*	*27*	*29*	*24*
	2	*21*	∞	*19*	*26*	*31*	*24*
From city	*3*	*26*	*28*	∞	*15*	*36*	*26*
	4	*25*	*16*	*25*	∞	*23*	*18*
	5	*23*	*20*	*23*	*31*	∞	*10*
	6	*27*	*18*	*12*	*35*	*16*	∞

The sales man can visit each of the cities once and only once. Determine the optimum sequence he should follow to minimize the total distance travelled. What as the total distance travelled?

Solution:

Following the usual procedure of assignment algorithms, we obtain Table 12.1 showing an optimum solution indicated by encircled zeros.

This table provides the optimum assignment schedule

1 → 3 → 4 → 2 → 1 and 5 → 6 → 5

with distance according to this optimum route being 101.

Table

	1	2	3	4	5	6
1	∞	0	0	7	2	14
2	0	∞	0	10	8	8
3	6	13	∞	0	14	11
4	4	0	6	∞	0	2
5	8	30	10	21	∞	0
6	13	6	0	26	0	∞

This table does not provide the solution to the travelling salesman problem, as it gives 1 → 3, 3 → 4, 4 → 2, 2 → 1, while city 2 is not allowed to follow city 1 unless city 5 and city 6 processed.

Now we try to find the next best solution which also satisfies this extra restriction. The next minimum (non-zero) element in the matrix is 2. Therefore, we try to bring 2 into the solution. But the element 2 occurs at two places. We shall consider both the cases separately until the acceptable solution is attained,

	1	2	3	4	5	6
1	∞	0	0	7	(2)	14
2	0	∞	0	10	8	8
3	6	13	∞	0	14	11
4	4	0	6	∞	0	2
5	8	30	10	21	∞	0
6	13	6	0	26	0	∞

We start with the element (1, 5) and make the assignment in this cell instead of zero assignment in the cell (1, 3). After making this assignment we see that no other assignment can be made in the first row and third column, and thus the resulting feasible solution will be

1 → 5, 5 → 6, 6 → 3,

3 → 4, 4 → 2, 2 → 1.

The selected elements for this solution are encircled in Table 12.2. The cost corresponding to this feasible solution is 2.

Again, if we make assignment in the cell (4, 6) instead of (4, 2) then no feasible solution is available in terms of zeros. Hence the best programme is

1 → 5 → 6 → 3 → 4 → 2 → 1.

The total set-up cost according to this route comes out to be 103.

Solve the travelling salesman problem given by the following data:

$c_{12} = 20,\ c_{13} = 4,\ c_{14} = 10,$

$c_{23} = 5,\ c_{24} = 6,\ c_{25} = 10,$

$c_{35} = 6$ and $c_{45} = 20$,

where $c_{ij} = c_{ij}$

and there is no route between cities i and j if a value for c_{ij} is not shown above.

Given the following matrix of set-up cost, show how to sequence production so as to minimize set-up cost per cycle:

		To				
		A	B	C	D	E
	A	∞	2	5	7	1
	B	6	∞	3	8	2
From	C	8	7	∞	4	7
	D	12	4	6	∞	5
	E	1	3	2	8	∞

A medical representative has to visit five stations A, B, C, D, and E. He does not want to visit any station twice before completing his four of all the stations, and wishes to return to the starting station. Costs of going from one stations to another are given below. Determine the optimal route:

	A	B	C	D	E
A	∞	2	4	7	1
B	5	∞	2	8	2
C	7	6	∞	4	6
D	10	3	5	∞	4
E	1	2	2	8	∞

EXERCISE

1. Write a short note on travelling salesman problem.
2. State the travelling salesman problem and formulate it as an assign ment problem.
3. How can the travelling select man problem be solved using assignment algorithm.

4. Solve the 'travelling salesman problem' given by the following data:

$c_{12} = 4$, $c_{13} = 7$, $c_{14} = 3$, $c_{23} = 6$, $c_{24} = 3$ and $c_{34} = 7$, where $c_{ij} = c_{ji}$.

5. Solve the travelling salesman problem in the matrix shown below:

	1	2	3	4	5
1	∞	6	12	6	4
2	6	∞	10	5	4
3	8	7	∞	11	3
4	5	4	11	∞	5
5	5	2	7	8	∞

6. A salesman has to visit five cities A, B, C, D and E. The distances (in hundred miles) between the five fifties are as follows:

		To				
		A	B	C	D	E
	A	–	7	6	8	4
	B	7	–	8	5	6
From	C	6	8	–	9	7
	D	8	5	9	–	8
	E	4	6	7	8	–

If the salesman starts from city A and has to come back to city A, which route should he select so that the total distance travelled is minimum.

7. Solve the following travelling salesman problem:

		To					
		1	2	3	4	5	6
	1	∞	20	23	27	29	34
	2	21	∞	19	26	31	24
From	3	26	28	∞	15	36	26
	4	25	16	25	∞	23	18
	5	23	40	23	31	∞	10
	6	27	18	12	35	16	∞

ANSWERS

4. 1 → 3 → 2 → 4 → 1; min. cost = 19.
5. 3 → 5 → 2 → 4 → 1 → 3; min. cost = 27.
6. A → E → B → D → C → A; min. distance = 30 hundred miles.
7. 1 → 5 → 6 → 3 → 4 → 2 → 1; min. cost = 103.

EXERCISE
(Objective Questions)

Fill in the Blanks:

Fill in the blanks" ... "so that the following statements are complete and correct.

1. The problems where the objective is to assign a number of origins to the equal number of destinations at a ... cost are called 'Assignment problem'.
2. In an assignment problem if x_{ij} denotes that the ith person is to be assigned the jth job then $\sum_{i=1}^{n} x_{ij}$ = ... and $\sum_{j=1}^{n} x_{ij}$ = ...
3. In an assignment problem with cost (c_{ij}), if all $c_{ij} \geq 0$, then a feasible solution (x_{ij}) which satisfies $\sum_{j=1}^{n}\sum_{j=1}^{n} c_{ij}x_{ij}$ = ... , is optimal for the problem.
4. In travelling salesman problem the element of the leading diagonal of the cost matrix are taken to be ...
5. If the cost matrix of an assignment problem is not a square matrix (number of sources is not equal to the number of destinations), the assignment problem is called an ... assignment problem.

Multiple Choice Questions:

Indicate the correct answer for each question by writing the corresponding letter from (a), (b), (c) and (d).

6. The complete optimal assignment is obtained if in the reduced cost matrix of order n the number of marked '□' zeros is

   (a) less than n

(b) greater than n

(c) exactly n

(d) none of these.

7. An optimal assignment exist if the total reduced cost of the assignment is

(a) zero

(b) one

(c) two

(d) none of these.

8. In an unbalanced assignment problem to form a square matrix fictitious rows or columns are added in the matrix with costs

(a) 1 (b) 0

(c) ∞ (d) none of these.

9. If salesman wants to visit n cities then the number of possible routes is

(a) n! (b) (n − 1)!

(c) n (d) none of these.

10. In the process of drawing minimum number of lines to cover all the zeros of the reduced matrix we draw lines through

(a) marked columns

(b) unmarked rows

(c) unmarked row and marked columns

(d) none of these.

True or False:

Write 'T' for true and 'F' for false statement.

11. If in an assignment problem, a constant is added or subtracted to every element of a row (or column) of the cost matrix $[c_{ij}]$, then an assignment which minimizes the total cost for one matrix, also minimizes the total cost for the other matrix.

12. For solving an assignment problem we modify the cost matrix by creating zeros in it.

13. If there is no solution to the travelling salesman problem among zeros then the best solution lies in assigning the greatest element of the reduced matrix in place of zero.

14. The procedure of subtracting the minimum element not covered by any line, from all the uncovered elements and adding the same element to all the elements lying at the intersection of two lines results in a matrix with different optimal assignments as the original matrix.

ANSWERS

1. Minimum. **2.** 1;1. **3.** 0. **4.** ∞. **5.** unbalanced.

6. (c). **7.** (a). **8.** (b). **9.** (b). **10.** (c). **11.** T. **12.** T.

13. F. **14.** F.

DUALITY

INTRODUCTION

The term duality is that every linear programming is associated with another linear programming problem called dual of problem. The original (given) problem is called the primal problem while the other is called its dual problem. It is important in general, that either problem can considered as primal and other as its dual. Thus, the two problems constitute the pair of dual problems.

DUALITY IN LINEAR PROGRAMMING

To understand the concept of duality more clearly, consider the following diet problem:

The amount of two vitamins A and B per unit present in two different foods A_1 and A_2 respectively are given below.

Vitamin	*Food*		*Minimum daily requirement*
	A_1	A_2	
A	6	9	60
B	4	13	108
Cost (per unit)	Rs. 12	Rs. 18	

The objective of the diet problem is to ascertain the quantities of foods A_1 and A_2 that should be eaten to meet the minimum daily requirements of vitamins A and B at a minimum cost.

Let x_1 and x_2 be the number of units of foods A_1 and A_2 to be purchased respectively. Then the problem is to find the values of x_1 and x_2 which minimize

$$Z_x = 12x_1 + 18x_2$$

subject to the constraints

$$6x_1 + 9x_2 \geq 60$$

$$4x_1 + 13x_2 \geq 108,$$

$$x_1, x_2, \geq 0.$$

We shall consider this linear programming problem as the primal problem. Now we shall formulate the dual corresponding to this primal.

Suppose a wholesale dealer sells two vitamins A and B. Customers purchase the two vitamins from him in the form of two foods A_1 and A_2 (as given in the above). The foods A_1 and A_2 have their market value only because of their vitamin contents. Now the dealer has to fix-up the maximum per unit selling prices for the two vitamins A and B in such a way that the resulting prices of the foods A_1 and A_2 do not exceed their existing market prices.

Suppose the dealer decides to fix-up the two prices y_1 and y_2 respectively. Then the problem is to determine the values of y_1 and Y_2 which maximize

$$Zy = 60y_1 + 108y_2$$

subject to

$$6y_1 + 4y_2 \leq 12$$

$$9y_1 + 12y_2 \leq 18,$$

$$y_1, y_2 \geq 0.$$

Observing the above primal and the dual a little closely, we find that

(i) Primal is a minimization problem while the dual is a maximization problem.

(ii) The constraint values 60, 108 of the primal have become the coefficients of the dual variables y_1 and y_2 in the objective function of the dual in that order. The coefficients of the variables x_1 and x_2 in the objective function of the primal have become the constraint values in the dual.

(iii) The constraint coefficient matrix of the dual is the transpose of the constraint coefficient matrix of the primal.

(iv) The direction of inequalities in the dual is the reverse of that in the primal.

We can represent the primal-dual relationship in the following form:

Primal	*Dual*
Minimize	Maximize
$Z_x = \overset{c}{[12\ 18]} \overset{x}{\begin{bmatrix} x_1 \\ x_2 \end{bmatrix}}$	$Z_y = \overset{b'}{[60\ 108]} \overset{y}{\begin{bmatrix} y_1 \\ y_2 \end{bmatrix}}$
subject to	subject to
$\overset{A}{\begin{bmatrix} 6 & 9 \\ 4 & 13 \end{bmatrix}} \overset{x}{\begin{bmatrix} x_1 \\ x_2 \end{bmatrix}} \geq \overset{b}{\begin{bmatrix} 60 \\ 108 \end{bmatrix}}$	$\overset{A'}{\begin{bmatrix} 6 & 4 \\ 9 & 13 \end{bmatrix}} \cdot \overset{y}{\begin{bmatrix} y_1 \\ y_2 \end{bmatrix}} \leq \overset{c'}{\begin{bmatrix} 12 \\ 18 \end{bmatrix}}$
$x_1, x_2 \geq 0.$	$y_1, y_2 \geq 0.$

SYMMETRIC DUAL PROBLEMS

A symmetric relation between a primal and its dual problem exists.

Primal Problem

Consider a L.P. problem.

$$\begin{aligned} \text{Mini} \quad & Z_p = c_1 x_1 + c_2 x_2 + \ldots + c_n x_n \\ \text{s.t.} \quad & a_{11} x_1 + a_{12} x_2 + \ldots + a_{1n} x_n \geq b_1 \\ & a_{21} x_1 + a_{22} x_2 + \ldots + a_{2n} x_n \geq b_2 \\ & a_{m1} x_1 + a_{m2} x_2 + \ldots + a_{mn} x_n \geq b_m \\ \text{and} \quad & x_1, x_2, x_3, \ldots, x_n \geq 0. \end{aligned}$$

The dual problem of above L.P. problem is obtained by (i) transporting the coefficient matrix. (ii) Interchanging the role of constant terms and the coefficients of the objective function. (iii) Changing the direction of inequalities. (iv) Maximizing instead of minimizing.

The dual problem is as follows.

$$\begin{aligned} \text{Max.} \quad & Z_D = b_1 w_1 + b_2 w_2 + \ldots + b_m w_m \\ \text{s.t.} \quad & a_{11} w_1 + a_{21} w_2 + \ldots + a_{m1} w_m \leq c_1 \\ & a_{21} w_1 + a_{22} + \ldots + a_{m2} w_m \leq c_2 \\ & a_{1n} w_1 + a_{2n} w_2 + \ldots + a_{mn} w_m \leq c_n \\ \text{and} \quad & w_1, w_2, \ldots, w_m \geq 0. \end{aligned}$$

In matrix form the primal and dual problems can be written as

Primal Problem: Find the column vector x which

Mini. $Z_p = c\,x$

s.t. $Ax \geq b$

and $x \geq 0.$

Dual Problem : Find a column vector w which

Maxi $Z_D = b'\,w$

s.t. $A'\,w \leq c',$

and $w \geq 0.$

The above primal and dual problem is called symmetric dual problems.

UNSYMMETRICAL DUAL PROBLEM

Primal Problem. Find a column vector x which

Mini. $Z_p = c\,x$

s.t. $A\,x = b,\ x \geq 0$

Dual problem. Find a column vector w which

Max. $Z_D = b'\,w$

s.t. $A'\,w \leq c'.$

The Dual of Mixed System : If a system consist of a monitor of equations, inequalities (in either direction), non-negative variable, or unrestricted.

MATRIX FORM OF SYMMETRIC PRIMAL-DUAL PROBLEM

Primal Problem. Find a column vector $\mathbf{x} \in R^n$ which maximizes

$$Z_x = \mathbf{c\,x},\ \mathbf{c} \in R^n$$

subject to $\mathbf{A\,x} \leq \mathbf{b},\ \mathbf{b} \in R^m,$

$x \geq 0$ and A is an $m \times n$ real matrix.

Dual Problem. Find a column vector $\mathbf{w} \in R^m$ which minimizes

$$Z_w = \mathbf{b'\,w}$$

subject to $\mathbf{A'\,w} \geq \mathbf{c'}$

$\mathbf{w} \geq 0$, $\mathbf{A'}$, $\mathbf{b'}$, $\mathbf{c'}$ are the transposes of $\mathbf{A}$, $\mathbf{b}$ and $\mathbf{c}$ respectively.

Example 1:

Consider the symmetric primal problem

Max. $Z_x = 5x_1 + 9x_2$

subject to $x_1 \leq 6$

$$x_1 + x_2 \le 13$$
$$x_2 \le 8,$$
$$x_1, x_2 \ge 0.$$

Solution:

The corresponding dual problem is

Min. $Z_w = 6w_1 + 13w_2 + 8w_3$

subject to $w_1 + w_2 \ge 5$

$$w_2 + w_3 \ge 9,$$
$$w_1, w_2, w_3 \ge 0.$$

Note: The following table a simple and convenient method to remember the primal dual relationship:

$(x_1, x_2, \ldots, x_n)$ Min.

$$\begin{bmatrix} w_1 \\ w_2 \\ \vdots \\ w_m \end{bmatrix} \begin{bmatrix} a_{11} & a_{12} & \cdots & a_{1n} \\ a_{21} & a_{22} & \cdots & a_{2n} \\ \cdots & \cdots & \cdots & \cdots \\ a_{m1} & a_{m2} & \cdots & a_{mn} \end{bmatrix} \pounds \begin{bmatrix} b_1 \\ b_2 \\ \vdots \\ b_m \end{bmatrix}$$

Max. $(c_1, c_2, \ldots, c_n)$

Variables then it is always possible to obtain the dual of this system. In fact every equation is equal to two inequalities one with less than and second with greater than inequality and every unrestricted variable is equal to difference of two positive variable. A constraint with $\le$ sign can be converted into a constraint with $\ge$ sign by multiply by -1.

Example 2:

Write the dual of the problem.

Maxi. $Z = 13x_1 + x_2$

s.t. $12x_1 + 3x_2 \le 20$

$$4x_1 + x_2 \le 35.$$

Solution:

In matrix form the given problem can be written as

Max. $Z = (13, 1)(x_1, x_2)$

$= cx$

s.t. $\begin{bmatrix} 12 & 3 \\ 4 & 1 \end{bmatrix} \begin{bmatrix} x_1 \\ x_2 \end{bmatrix} \le \begin{bmatrix} 20 \\ 35 \end{bmatrix}$

or $\quad ax \leq b.$

The dual of the given problem is

Mini. $\quad Z_D = b' w$

$= (20, 35) (w_1, w_2)$

Mini. $\quad Z_D = 20w_1 + 35w_2$

s.t. $\quad A' w \geq c'$

$$\begin{bmatrix} 12 & 4 \\ 3 & 1 \end{bmatrix} \begin{bmatrix} w_1 \\ w_2 \end{bmatrix} \geq \begin{bmatrix} 13 \\ 1 \end{bmatrix}$$

$12w_1 + 4w_2 \geq 13$

$3w_1 + w_2 \geq 1$

$w_1, w_2 \geq 0.$

Example 3(a):

Find the dual of the L.P. problem

Mini $\quad Z = 4x_1 + 6x_2 + 18x_3$

s.t. $\quad x_1 + 3x_3 \geq 3$

$x_2 + 2x_3 \geq 5$

$x_1, x_2, x_3 \geq 0.$

Solution:

In matrix from the given problem can be written as

Mini. $\quad Z = 4, 6, 18) (x_1, x_2, x_3)$

$= c\, x$

s.t. $$\begin{bmatrix} 1 & 0 & 3 \\ 0 & 1 & 2 \end{bmatrix} \begin{bmatrix} x_1 \\ x_2 \\ x_3 \end{bmatrix} \geq \begin{bmatrix} 3 \\ 5 \end{bmatrix}$$

or $AX \geq b$

The dual of the given problem is

Max. $Z_D = b' w$

$= (3, 5) (w_1, w_2)$

$= 3w_1 + 5w_2$

Max. $Z_D = 3w_1 + 5w_2$

s.t. $\quad A' w \leq c'$

$$\begin{bmatrix} 1 & 0 \\ 0 & 1 \\ 3 & 2 \end{bmatrix} \begin{bmatrix} w_1 \\ w_2 \end{bmatrix} \le \begin{bmatrix} 4 \\ 6 \\ 18 \end{bmatrix}$$

or $1.w_1 + 0w_2 \le 4$

$ow_1 + 1w_2 \le 6$

$3w_1 + 2w_2 \le 18$

$w_1, w_2 \ge 0.$

Example 3(b):

Write the dual of the L.P. problem

Max. $Z = 2x_1 + x_2$

s.t. $x_1 + 2x_2 \le 10$

$x_1 + x_2 \le 6$

$x_1 - x_2 \le 2$

$x_1 - 2x_2 \le 1$

$x_1, x_2, x_3 \ge 0.$

Solution:

In matrix from the given problem can be written as

Max. $Z = (2, 1)(x_1, x_2) = c\,x$

s.t. $$\begin{bmatrix} 1 & 2 \\ 1 & 1 \\ 1 & -1 \\ 1 & -2 \end{bmatrix} \begin{bmatrix} x_1 \\ x_2 \end{bmatrix} \le \begin{bmatrix} 10 \\ 6 \\ 2 \\ 1 \end{bmatrix}$$

or $AX \le b$

The dual of the given problem is

Mini. $Z_D = b'\,w$

$= [10, 6, 2, 1]\,[w_1, w_2, w_3, w_4]$

$Z_D = 10w_1 + 6w_2 + 2w_3 + w_4$

s.t. $A'\,w \ge c'$

or $$\begin{bmatrix} 1 & 1 & 1 & 1 \\ 2 & 1 & -1 & -2 \end{bmatrix} \begin{bmatrix} w_1 \\ w_2 \\ w_3 \\ w_4 \end{bmatrix} \ge \begin{bmatrix} 2 \\ 1 \end{bmatrix}$$

$w_1 + w_2 + w_3 + w_4 \ge 2$

$$2w_1 + w_2 - w_3 - 2w_4 \geq 1$$

$$w_1, w_2, w_3, w_4 \geq 0$$

Example 4:

The dual of L.P. problem.

Max. $Z = 2x_1 + 2x_2$

s.t. $2x_1 - 3x_2 \leq 3$

$4x_1 + x_2 \leq -4$

$x_1, x_2 \geq 0.$

Solution:

In matrix from the given problem can be written as

Max. $Z = (2, 2)\,(x_1, x_2)$

$= c\,x$

s.t. $\begin{bmatrix} 2 & -3 \\ 4 & 1 \end{bmatrix} \begin{bmatrix} x_1 \\ x_2 \end{bmatrix} \leq \begin{bmatrix} 3 \\ -4 \end{bmatrix}$

or $A\,X \leq b$

The dual of the given problem is

Mini $Z_D = b'\,w$

$= (3, -4)\,(w_1, w_2)$

$Z_D = 3w_1 - 4w_2$

s.t. $A'\,w \geq c'$

$$\begin{bmatrix} 2 & 4 \\ -3 & 1 \end{bmatrix} \begin{bmatrix} w_1 \\ w_2 \end{bmatrix} \geq \begin{bmatrix} 2 \\ 2 \end{bmatrix}$$

$$2w_1 + 4w_2 \geq 2$$

$$-3w_1 + w_2 \geq 2$$

$$w_1, w_2 \geq 0$$

Example 5:

Find the dual of the following L.P.P.

Max. $Z = 6x_1 + 5x_2 + 10x_3$

s.t. $4x_1 + 5x_2 + 7x_3 \leq 5$

$3x_1 + 0x_2 + 7x_3 \leq 10$

$$2x_1 + x_2 + 8x_3 \leq 20$$
$$0x_1 + 2x_2 + 9x_3 \geq 5$$
$$x_1, x_3 \geq 0$$

x_2 an restricted in sign.

Solution:

First of all we shall write the given problem in standard form as follows.

(i) Since it is a maximization problem, all the constraints must contain the sign $\leq$.

Therefore we multiply the fourth constraints by -1 we get

$$-2x_2 - 9x_3 \leq -5.$$

(ii) The variable x_2 is unrestricted in sign

$\therefore$ we write $x_2 = x_2 - x_2''$ where $x_2', x_2'' \geq 0$

So first constraint is equivalent to

$$4x_1 + 5(x_2' - x_2'') + 7x_3 \leq 5$$

The third constraint can be written as

$$2x_1 + (x_2' - x_2'') + 8x_3 \leq 20$$

and fourth constraint is equivalent to

$$2(x_2' - x_2'') + 9x_3 \geq 5$$

or $$-2x_2' + 2x_2'' - 9x_3 \leq -5$$

Thus the given problem is standard primal form is

$$\text{Max. } Z = 6x_1 + 5(x_2' - x_2'') + 10x_3$$
$$= (6, 5, -5, 10)\,[x_1, x_2,' x_2,'' x_3]$$
$$= c\,x$$

s.t. $$4x_1 + 5(x_2' - x_2'') + 7x_3 \leq 5$$
$$3x_1 + 7x_3 \leq 10$$
$$2x_1 + x_2' - x_2'' + 8x_3 \leq 20$$
$$-2x_2' + 2x_2'' - 9x_3 \leq -5$$

or $$\begin{bmatrix} 4 & 5 & -5 & 7 \\ 3 & 0 & 0 & 7 \\ 2 & 1 & -1 & 8 \\ 0 & -2 & 2 & -9 \end{bmatrix} \begin{bmatrix} x_1 \\ x_2' \\ x_2'' \\ x_3 \end{bmatrix} \leq \begin{bmatrix} 5 \\ 10 \\ 20 \\ -5 \end{bmatrix}$$

$$A\,x \leq b$$

and $x_1, x_2', x_2'', x_3 \geq 0$

The dual of the given problem is

Mini. $Z_D = b' w$

$$= [5, 10, 20, -5]\,[w_1, w_2, w_3, w_4]$$

$$Z_D = 5w_1 + 10w_2 + 20w_3 - 5w_4$$

s.t. $A' w \leq c'$

$$\begin{bmatrix} 4 & 3 & 2 & 0 \\ 5 & 0 & 1 & -2 \\ -5 & 0 & -1 & 2 \\ 7 & 7 & 8 & -9 \end{bmatrix} \begin{bmatrix} w_1 \\ w_2 \\ w_3 \\ w_4 \end{bmatrix} \leq \begin{bmatrix} 6 \\ 5 \\ -5 \\ 10 \end{bmatrix}$$

$$4w_1 + 3w_2 + 2w_3 + 0w_4 \geq 6$$

$$5w_1 + 0w_2 + 1w_3 - 2w_4 \geq 5$$

$$-5w_1 + 0w_2 - 1w_3 + 2w_4 \geq -5$$

$$7w_1 + 7w_2 + 8w_3 - 9w_4 \geq 10$$

$$w_1, w_2, w_4 \geq 0$$

and w_3 is unrestricted.

Example 6:

Find the dual of following L.P.P.

Mini. $Z = x_1 + x_2 + x_3$

s.t. $x_1 - 3x_2 + 4x_3 = 5$

$x_1 - 2x_2 \leq 3$

$2x_2 - x_3 \geq 4$

$x_1, x_2 \geq 0$, x_3 *is unrestricted in sign.*

Solution:

First of all we shall write the given problem in standard form as follows.

(i) It is a minimization problem, all the constraints must contain the sign $\geq$.

Therefore we multiply the second constraints by -1 we get

$$-x_1 + 2x_2 \geq -3.$$

(ii) The variable x_3 is unrestricted in sign

$\therefore$ we write $x_3 = x_3' - x_3''$ where $x_3', x_3'' \geq 0$

So first constraint is equivalent to

$$x_1 - 3x_2 + 4\,(x_3' - x_3'') \geq 5 - \quad \text{(I)}$$

$$x_1 - 3x_2 + 4\,(x_3' - x_3'') \leq 5 - \quad \text{(II)}$$

Multiply above II by –1

$$-x_1 + 3x_2 - 4\,(x_3' - x_3'') \geq -5$$

Thus the given problem is standard primal form is

Mini. $Z = x_1 + x_2 + x_3' - x_3''$

$= (1, 1, 1, -1)\,[x_1, x_2, x_3', x_3'']$

$= c\,x$

s.t. $x_1 - 3x_2 + 4x_3' - 4x_3'' \geq 5$

$-x_1 + 3x_2 - 4x_3' + 4x_3'' \geq -5$

$-x_1 + 2x_2 + 0x_3' - 0x_3'' \geq -3$

$0x_1 + 2x_2 - x_3' + x_3'' \geq 4$

or
$$\begin{bmatrix} 1 & -3 & 4 & -4 \\ -1 & 3 & -4 & 4 \\ -1 & 2 & 0 & 0 \\ 0 & 2 & -1 & 1 \end{bmatrix} \begin{bmatrix} x_1 \\ x_2 \\ x_3 \\ x_3 \end{bmatrix} \geq \begin{bmatrix} 5 \\ -5 \\ -3 \\ 4 \end{bmatrix}$$

or $A\,x \geq b$

and $x_1, x_2, x_3', x_3'' \geq 0$

∴ The dual of the given problem is

Max. $Z_D = b'\,w$

$= (5, -5, -3, 4)\,(w_1', w_1'', w_2, w_3)$

$= 5w_1' - 5w_1'' - 3w_2 + 4w_3$

s.t. $A'w \geq c'$

$$\begin{bmatrix} 1 & -1 & -1 & 0 \\ -3 & 3 & 2 & 2 \\ 4 & -4 & 0 & -1 \\ -4 & 4 & 0 & 1 \end{bmatrix} \begin{bmatrix} w_1' \\ w_1'' \\ w_2 \\ w_3 \end{bmatrix} \leq \begin{bmatrix} 1 \\ 1 \\ 1 \\ -1 \end{bmatrix}$$

or $w_1' - w_1'' - w_2 + 0.w_3 \leq 1$

$-3w_1' + 3w_1'' + 2w_2 + 2w_3 \leq 1$

$4w_1' - 4w_1'' + 0.w_2 - w_3 \leq 1$

$-4w_1' + 4w_1'' + 0.w_2 + w_3 \leq -1$

and $w_1', w_1'', w_2, w_3 \geq 0$

writing $w_1' - w_1'' = w_1$, the dual problem is also written as

Max. $Z_D = 5w_1 - 3w_2 - 4x_3$

s.t. $w_1 - w_2 \leq 1$

$-3w_1 + 2w_2 + 2w_3 \leq 1$

$4w_1 - w_3 = 1$

$w_2, w_3 \leq 0$, w_1 unrestricted in sign.

EXERCISE

Write the dual of the following L.P. problem.

1. Mini. $Z = 3x_1 + x_2$

s.t. $2x_1 + 3x_2 \geq 2$

$x_1 + x_2 \geq 1$

and $x_1, x_2 \geq 0$.

2. Max. $Z = 3x_1 + 5x_2 + 4x_3$

s.t. $2x_1 + 3x_2 \leq 8$

$2x_2 + 5x_3 \leq 10$

$3x_1 + 2x_2 + 4x_3 \leq 15$

and $x_1, x_2, x_3 \geq 0$.

3. Mini $Z = 15x_1 + 10x_2$

s.t. $3x_1 + 5x_2 \geq 5$

$5x_1 + 2x_2 \geq 3$

$x_1, x_2 \geq 0$

4. Max. $Z = 3x_1 - 2x_2$

$x_1 \leq 4$

$x_2 \leq 6$

$x_1 + x_2 \leq 5$

$-x_2 \leq -1$

$x_1, x_2 \geq 0$.

5. Max. $Z = x_1 + 3x_2$

s.t. $3x_1 + 2x_2 \leq 6$

$3x_1 + 2x_2 = 4$

$x_1, x_2 \geq 0$.

ANSWERS

1. Max. $Z_D = 2w_1 + w_2$

s.t. $2w_1 + w_2 \leq 3$

$3w_1 + w_2 \leq 1$

and $w_1, w_2 \geq 0.$

2. Mini. $Z_D = 8w_1 + 10w_2 + 15w_3$

$2w_1 + 3w_3 \geq 3$

$3w_1 + 2w_2 + 2w_3 \geq 5$

$5w_2 + 4w_2 \geq 4$

and $w_1, w_2, w_3 \geq 0.$

3. Max $Z_D = 5w_1 + 3w_2$

s.t. $3w_1 + 5w_2 \leq 15$

$5w_1 + 2w_2 \leq 10$

$w_1, w_2 \geq 0.$

4. Mini $Z_D = 4w_1 + 6w_2 + 5w_3 - w_4$

s.t. $w_1 + w_3 \geq 3$

$w_2 + w_3 - w_4 \geq -2$

$w_1, w_2, w_3, w_4 \geq 0.$

5. Mini. $Z_D = 6w_1 + 4w_2$

s.t. $3w_1 + 3w_2 \geq 1$

$2w_1 + w_2 \geq 3$

and $w_1 \geq 0$

w_2 unrestricted in sign.

Theorem 1:

Prove that dual of a dual is primal problem.

Proof:

Let primal problem is

$$\left.\begin{aligned} &\text{Max. } Z = c\,x \\ &\text{s.t. } \quad A\,x \leq b \\ &\qquad x \geq 0 \end{aligned}\right\} \qquad ...(1)$$

Dual is Mini. $Z_{D1} = b'w$
s.t. $A'w \geq c'$
and $w \geq 0$ ...(2)

Thus dual (2) can be written as

$$\left.\begin{aligned} \text{Max} \quad (-Z_{D1}) &= -b'w \\ &= d\,w \\ \text{s.t.} \quad -A'w &\leq -c' \\ w &\geq 0 \end{aligned}\right\} \qquad [\text{Put } -b' = d] \quad ...(3)$$

Dual of the 3rd constraint is

$$\left.\begin{aligned} \text{Mini} \quad (-Z_{D1}) &= (-c')'\,y \\ \text{s.t.} \quad (-A')'\,y &\geq (-b')' \\ y &\geq 0 \end{aligned}\right\} \quad ...(4)$$

The dual (4) can be written as

$$\left.\begin{aligned} \text{Max.} \quad Z_{D2} &= c\,y \\ \text{s.t.} \quad A\,y &\leq b \\ y &\geq 0 \end{aligned}\right\}$$

which is similar to (1) *i.e.*, the primal.

Theorem 2:

$$\left.\begin{aligned} \text{If Max.} \quad Z_P &= c\,x \\ \text{s.t.} \quad A\,x &\leq b \\ x &\geq 0 \end{aligned}\right\} \text{ is primal}$$

$$\left.\begin{aligned} \text{Mini.} \quad Z_D &= b'w \\ \text{s.t.} \quad A'w &\geq c' \\ w &\geq 0 \end{aligned}\right\} \text{ is dual}$$

Prove that $D_P \leq Z_D$ $\left\{\begin{array}{l} \forall\ x \text{ is feasible solution of primal.} \\ \forall\ w \text{ is feasible solution of dual.} \end{array}\right.$

Proof:

$$\left.\begin{aligned} \text{Primal Max.} \quad Z_P &= c\,x \\ \text{s.t.} \quad A\,x &\leq b \\ x &\geq 0 \end{aligned}\right\} \quad ...(1)$$

$$\left.\begin{array}{ll}\text{Dual Mini.} & Z_D = b'\,w \\ \text{s.t.} & A'\,w \geq c' \\ & w \geq \end{array}\right\} \quad \ldots(2)$$

x and w be any feasible solution to the primal (1) and dual problem (2) respectively.

If we multiply the w' a row matrix in the constraint of primal

$$w'\,A\,x \leq w'\,b \quad \ldots(3)$$

and multiply the x' a column matrix in the constraint of dual

$$x'\,A'\,w \leq x'\,c' \quad \ldots(4)$$

equation (4) can be written as

$$(A\,x)'\,w \geq (c\,x)'$$

Taking transpose both side

$[(A\,x)'\,w]'\,A'\,x' = (x\,A)'\,[(c\,x)]'$ Matrix problem $A'\,x' = (x\,A)'$

$$(A\,x)'' = A\,x$$

$$w'\,(A\,x)'' \geq (c\,x)''$$

$$w'\,A\,x \geq c\,x \quad \ldots(5)$$

from (1) and (5)

$$w'\,A\,x \geq Z_P \quad \ldots(6)$$

from (3) and (6)

$$w'\,b \geq w'\,A\,x \geq Z_P$$

$$(b'\,w)' \geq Z_P$$

(Since $(b'\,w)' = b'\,w$, as $b'\,w$ is a scalar value, not a matrix)

So $\quad b'\,w \geq Z_P \quad \ldots(7)$

from (2) and (7)

$$Z_D \geq Z_P.$$

Theorem 3:

If $\overline{x}$ and $\overline{w}$ are feasible solution of primal & usual respectively such that $c\,\overline{x} = b'\,\overline{w}$ (*i.e.*, $Z_P = Z_D$) then x and w are optimum solution of the respective problem respectively *i.e.*, $\overline{x}$ is max and $\overline{w}$ is mini problem

Max $\quad Z_P = c\,x$

s.t. $\quad A\,x \leq b,\ x \geq 0$

and its corresponding dual

Mini. $Z_D = b'w$

$A'w \geq c'\ w \geq 0.$

Or Alternative Statement

The necessary and sufficient condition for any L.P.P. and its dual to have optimal solution is that both have feasible solution.

Proof:

Let $\overline{x}$ is any feasible solution to the primal

Max. $Z_P = c\,\overline{x}$, ...(1)

s.t. $A x \leq b,\ x \geq 0$

and w is any feasible solution of dual

Mini. $Z_D = b'w$

s.t. $A'w \geq c',\ w \geq 0$ ...(2)

then from theorem (2), we have

$c\,\overline{x} \leq b'w$ $(Z_P \leq Z_D)$

$b'\,\overline{w} = c\,x \leq b'w$ (as $c\,\overline{x} = b'w$ given)

i.e., $b'\,\overline{w} \leq b'w$

$\Rightarrow \overline{w}$ is mini solution of dual let $\overline{w}$ is given feasible solution of dual x is any feasible solution of primal

$c\,x \leq b'\,\overline{w} = c\,\overline{x}$

$\Rightarrow$ $c\,x \leq c\,\overline{x}$

$\Rightarrow$ $c\,\overline{x}$ is max value of the objective function

$\Rightarrow$ $\overline{x}$ is maximum solution of the primal problem.

Theorem 4:

If the L.P.P. mini. $Z_p = c\,x$ & & $A v \geq b$ has finites optimal solution than the dual max. $Z = b'w$ & $A'w \leq c'$ also has a finite maxi solution and if dual has a finite optimal solution then so as primal.

Proof:

Let the primal is

Mini $Z = c\,x$

& $a_{11}x_1 + a_{12}x_2 + \ldots + a_{1n}x_n \geq b_1$

$a_{m1}x_1 + a_{m2}x_2 + \ldots + a_{mn}x_n \geq b_m$

changing in equation

$a_{11}\, x_1 + a_{12}\, x_2 \ldots + a_{1n}\, x_n - x_{1s} = b_1$

$a_{m1}\, x_1 + a_{m2}\, x_2 + \ldots + a_{mn}\, x_n - x_{ms} = b_m$

The L.P.P. can be written as

$$A\,x - I\,X_s = b$$

Let x_B is a basic feasible solution which is optimal for the primal corresponding to base $B = (\alpha_1, \alpha_2, \ldots \alpha_m)$

$\Rightarrow x_B$ minimum solution of primal

$\Rightarrow$for this solution all $Z_j - c_j \leq 0$

$\Rightarrow c_B\, B^{-1}\, \alpha_j - c_j \leq 0$ for all j either original or for surplus variables

$\Rightarrow c_B\, B^{-1}\, \alpha_1 \leq c_1$

$c_B\, B^{-1}\, \alpha_2 \leq c_2$

$c_B\, B^{-1}\, \alpha_n \leq c_n$ *i.e.*, for original variable

$\Rightarrow (c_B\, b^{-1})\, (\alpha, \alpha, \ldots, \alpha_n \leq (c_1, c_2 - c_n)$

$\Rightarrow w_B'\, A \leq c$

Taking transport let $(c_B\, B^{-1} = w_B')$

$\Rightarrow A'\, w_B \leq c'$

$\Rightarrow w_B$ satisfy $A'\, w \leq c'$ *i.e.*, of dual

$\Rightarrow w_B$ is a dual solution.

I[st] we prove w_B is feasible solution IInd we prove w_B is optimum solution

$$w_B' = c_B\, (B^{-1})$$

Since $Z_J - c_j \leq 0$ for surplus variables also

$\Rightarrow c_B\, B^{-1}\, \alpha_{js} - 0 \leq 0$

$$\Rightarrow (w_{B1}, w_{B2}, \ldots, w_{Bm}) \begin{pmatrix} 0 \\ 0 \\ 0 \\ 1 \\ -1 \\ 0 \end{pmatrix}_{J^{\text{the}}\,\text{place}} \leq 0$$

$\Rightarrow - w_{BJ} \leq 0$ for $j = 1, 2, -m$

$\Rightarrow w_{BJ} \geq 0\ j = 1, 2, \ldots, m$

$\Rightarrow w_B = (c_B\, B^{-1})$ is a feasible solution of dual.

Part II.

Since $Z_P = (c_B\ B^{-1})\ b$

$= w_B'\ b$

$= b'\ w_B$

$= Z_D$

$\Rightarrow w_B = (c_B\ B^{-1})^1$ is max solution of dual so it is proved that if L.P.P. Min $Z_D = c\ x$ & $A\ x \geq b$ has an finite optimum solution then the dual max $Z_D = b'\ w$ and $A'\ w \leq c'$ also has a maximum solution.

Theorem 5:

If the primal problem has an unbounded solution, then the dual has either no solution or an unbounded solution.

Proof:

We shall prove this theorem by contradiction we suppose that primal has an unbounded solution, dual has a finite optimal solution. By theorem we have proved that dual of dual is primal and if primal has an optimal feasible solution then dual also has a finite optimal solution. Thus if we consider the dual as primal then its dual must have a finite optimal solution, which is a contradiction hence the dual has no finite optimal solution, and if it has it will be unbounded.

Theorem 6:

If any of the constraints in the primal is a strict equality, the corresponding dual variable is unrestricted in sign.

Proof:

Let L.P.P. is

Mini $Z_P = c\ x$

s.t. $\sum_{j=1}^{n} a_{ij} x_j \geq b_i\ \forall_i$

and $\sum_{j=1}^{n} a_{kj} x_j = b_k\ \forall_i \geq k$

$x_j \geq 0\ \forall_j$

In the standard primal form the above L.P.P. can be written as

Min $Z_P = c\ x$

s.t. $\sum_{j=1}^{n} a_{ij}x_j \geq b_i \ \forall_i \geq k$

and $\sum_{j=1}^{n} a_{kj}x_j \geq b_k$

$- \sum_{j=1}^{n} a_{kj}x_j \geq - b_k$

Now the dual of the above problem becomes

$$\text{Max. } Z_D = \sum_{i=1}^{m} b_i w_i + b_k (w_k' - w_k'')$$

$w_i \geq 0, \ i \neq k, \ w_k', \ w_k'' \geq 0$

Now writing $w_k'' - w_k'' = w_k$, the dual problem becomes

$$\text{Max. } Z_D = \sum_{i=1}^{m} b_i w_i$$

s.t. $\sum_{i=1}^{m} a_{ji}w_j \leq c_j \ \forall_j = 1, 2, \ldots n$

$w_j \geq 0 \ \forall_j \geq k$

obviously w_k is unrestricted in sign since $w_k >$ or < 0 according as $w_k' >$ or w_k''.

Theorem 7:

If any variable of the primal is unrestricted in sign, the corresponding constraint in the dual will be a stick equality.

Proof:

Let the kth variable x_k of the primal be unrestricted in sign.

Therefore replacing x_k by $x_k' - x_k''$ where x_k'' and x_k'' are non-negative variables, the primal can be written as

Max. $Z_D = c_1 x_1 + c_2 x_2 + \ldots + c_k (x_k' - x_k'') + c_{k+1} x_{k+1} + \ldots + c_n x_n$

s.t. $a_{11} x_1 + a_{12} x_2 + \ldots + a_{1k} (x_k' - x_k'') + a_{k+1} x_{k+1} + \ldots + a_{1n} x_n \leq b_1$

....

....

....

$a_{m1} x_1 + a_{m2} x_2 + \ldots + a_{mk} (x_k' - x_k'') + a_{m}, b_{11} x_{b+1} + a_{mn} x_n \leq b_m$

and $x_1, \ldots, x_{k-1}, x_k', x_b'', x_{k+1}, \ldots, x_n \geq 0$

The dual of the above problem is given by

Min. $Z_D = b_1 w_1 + b_2 w_2 + ... + b_m w_m$

and $\quad a_{11} w_1 + a_{21} w_2 + ... + a_{m1} w_m \geq c_1$

....

....

$$\left.\begin{array}{l} a_{12} w_1 + a_{22} w_2 + ... + a_{m2} w_m \geq c_2 \\ a_{1k} w_1 + a_{2k} w_2 + ... + a_{mk} w_m \geq c_k \end{array}\right\}$$

....

....

$a_{1k} w_1 - a_{2k} w_2 - ... - a_{mk} w_m \geq - c_k$

$a_{1n} w_1 + a_{2n} w + ... + a_{mn} w_m \geq c_n$

and $\quad w_1, w_2, ... , w_m \geq 0$

The two constraints under bracket {} in the above dual are equivalent to an equality

$$a_{1k} w_1 + a_{2k} w_2 + ... + a_{mk} w_m = c_k$$

i.e., the kth constraint in the dual is an equality.

Difference between Primal and Dual

Primal	*Dual*
1. Objective function Mini ZP.	Objective function max Z_D.
2. Requirement vector	Price vector
3. Coefficient matrix	Transpose of the coefficient matrix
4. Relation	Variable
5. i^{th} inequality	i^{th} variable $w_i \geq 0$
6. i^{th} constraint an equality.	i^{th} variable w_i unrestricted in sign.
7. constants with sign $\geq$	Constants with sign $\leq$
8. Variable	
9. i^{th} variable $x_i > 0$	i^{th} relation a strict equation.
10. i^{th} variable x_i unrestricted in sign	i^{th} constraint a strict equality
11. i^{th} slack variable positive	i^{th} variable zero
12. i^{th} variable zero	i^{th} surplus variable positive
13. Finite optimal solution	Finite optimal solution with equal optimal value of the objective function.
14. Unbounded solution	No solution or an unbounded solution.

Example 1:

Using the dual, solve the L.P.P.

Mini. $Z_P = 3x_1 + x_2$

s.t. $2x_1 + 3x_2 \geq 2$

$x_1 + x_2 \geq 1$

$x_1, x_2, \geq 0.$

Solution:

The dual of the given L.P.P. is

Max. $Z_D = 2w_1 + w_2$

s.t. $\left.\begin{array}{l} 2w_1 + w_2 \leq 3 \\ 3w_1 + w_2 \leq 1 \end{array}\right\}$

$w_1, w_2, \geq 0$

Introducing the slack variable w_{1s} and w_{2s} respectively. The dual problem can be written as

Max. $Z_D - 2w_1 + w_2$

s.t. $2w_1 + w_2 + w_{1s} = 3$

$3_{w1} = w_2 + w_{2s} = 1$

Taking $w_1 = 0 = w_2$, we have $w_{1s} = 3$ and $w_{2s} = 1$. Which is the starting basic feasible solution of the dual.

First Simplex Table

		c_J	2	1	0	0	w_B
B	c_B	w_B	Y_1	Y_2	Y_{1s}	Y_{2a}	Y_{i1}
x_{1s}	0	3	2	1	1	0	3/2
x_{2s}	0	1	[3]	1	0	1	1/3→mini
	Z_J ↑	$-c_J$	–2	–1	0	0 ↓	

incoming vector   outgoing vector

$\Delta_1 = Z_1 - c_1 = c_B Y_1 - c_1 = [(0 \times 2 = 0 \times 3) - 2] = -2$

$\Delta_2 = Z_2 - c_2 = c_B y_2 - c_2 = [0 \times 1 + 0 \times 1] - 1 = -1$

$\Delta_3 = Z_3 - c_3 = c_B y_3 - c_3 = 0$

$\Delta_4 = 0$

To find incoming vector

Since $\Delta_1 = -1$ is mini. of $\Delta_1, \Delta_2, \Delta_3, \Delta_4$. So y_1 is incoming vector 0.

To find outgoing vector

Since y_1 is incoming vector, so we find ratio Mini $\left\{\frac{w_B}{y_{i1}}\right\}$

$$= \text{Mini.} \left\{\frac{w_B}{y_{11}}, \frac{w_B}{y_{21}}\right\}$$

$$= \text{Mini} \quad \left\{\frac{3}{2}, \frac{1}{3}\right\}$$

$\frac{1}{3}$ is mini so x_{2s} is out going vector.

The key element is $y_{21} = 3$.

Firstly, we divide the second marked row of the first table by 3 to get the second row of the second table.

Now, Multiply the new second row by 2 (number in first row and marked column of the first table).

This will give

$$\frac{1}{3}\times 2 \qquad 1 \times 2 \qquad \frac{1}{3}\times 2 \qquad 0 \times 2 \qquad \frac{1}{3}\times 2$$

$$\frac{2}{3} \qquad 2 \qquad \frac{2}{3} \qquad 0 \qquad \frac{2}{3}.$$

Now subtracting these from the corresponding element of the first row of the first table. This will give

$$3-\frac{2}{3} \qquad 2-2 \qquad 1-\frac{2}{3} \qquad 1-0 \qquad 0-\frac{2}{3}$$

$$\frac{7}{3} \qquad 0 \qquad \frac{1}{3} \qquad 1 \qquad -\frac{2}{3}.$$

This is the first row of the second table.

Second Table

B	c_B	c_J / w_B	2 / Y_1	1 / Y_2	0 / Y_{1s}	0 / Y_{2s}	w_B / Y_{12}
x_{1s}	0	7/3	0	1/3	1	–2/3	7
x_1	2	1/3	1	[1/3]	0	1/3	1 mini→
			0 ↓ outgoing vector	–1/3	0 ↑ incoming vector	2/3	

$\Delta_1 = Z_1 - c_1 = 0$

$\Delta_2 = Z_2 - c_2 = -1/3$

$\Delta_3 = Z_3 - c_3 = 0$

$\Delta_4 = Z_4 - c_4 = 2/3$

To find incoming vector

Since $\Delta_2 = -\dfrac{1}{3}$ is mini of all. so y_2 is incoming vector

To find out vector.

Since y_2 is incoming vector. So we will find ratio Mini. $\left\{\dfrac{w_B}{y_{i2}}\right\}$

$$= \text{Mini}\left\{\frac{w_B}{y_{12}}, \frac{w}{y_{22}}\right\}$$

$$= \text{Mini}\left\{\frac{7\times3}{3\times1}, \frac{1\times3}{3\times1}\right\}$$

$$= \text{Mini}\ \{7, 1\}$$

1 is mini so x_1 is out going vector.

The key element is $y_{22} = 1/3$.

First we divide the second marked row of the second table by 1/3 to get the new second row of the third table.

Now multiply the new second row by 1/3 (number in the first row and marked column of the second table.

This will give

$$1\times\frac{1}{3} \qquad 3\times\frac{1}{3} \qquad 1\times\frac{1}{3} \qquad 0\times\frac{1}{3} \qquad 1\times\frac{1}{3}$$

$$\frac{1}{3} \qquad 1 \qquad \frac{1}{3} \qquad 0 \qquad \frac{1}{3}$$

Now subtracting these from the corresponding element of the first row of the second table. This will give

$$\frac{7}{3}-\frac{1}{3} \qquad 0-1 \qquad \frac{1}{3}-\frac{1}{3} \qquad 1-0 \qquad \frac{-2}{3}\frac{-1}{3}$$

$$2 \qquad -1 \qquad 0 \qquad 1 \qquad -1.$$

This is the first row of the third table.

Third Table

B	c_B	c_J / w_B	2 / Y_1	1 / Y_2	0 / Y_{1s}	0 / Y_{2s}
x_{1s}	0	2	–1	0	1	–1
x_2	1	1	3	1	0	1
		$Z_J - c_J$	1	0	0	1

Since all $Z_J - c_J \geq 0$, so this solution is optimal.

Optimal solution of the dual problem is

$$w_1 = 0$$

$$w_2 = 1$$

$$\text{Max} \quad Z_D = 1$$

solution of original problem

Form the find table solution of the primal is given by

$$x_1 = 0$$

$$x_2 = 1$$

$$\text{Mini. } Z_P = \text{Max } Z_D = 1$$

Example 2:

Using the dual solve the L.P.P.

$$\text{Mini} \quad Z_P = 10x_1 + 6x_2 + 2x_3$$

$$\text{s.t.} \quad -x_1 + x_2 + x_3 \geq 1$$

$$3x_1 + x_2 - x_3 \geq 2$$

$$x_1, x_2, x_3, x_4 \geq 0.$$

Solution:

The dual of this problem is given by

$$\text{Max} \quad Z_D = w_1 + 2w_2$$

$$\text{s.t.} \quad -w_1 + 3w_2 \leq 10$$

$$w_1 + w_2 \leq 6$$

$$w_1 - w_2 \leq 2$$

$$w_1, w_2 \geq 0.$$

Introducing the slack variable w_{1s}, w_{2s} and w_{3s} respectively. The dual problem can be written as

$$\text{Max} \quad Z_D = w_1 + 2w_2$$

$$\text{s.t.} \quad -w_1 + 3w_2 + w_{1s} = 10$$

$$w_1 + w_2 + w_{2s} = 6$$

$$w_1 - w_2 + w_{3s} = 2$$

Taking $w_1 = 0 = w_2$, we have $w_{1s} = 10$, $w_{2s} = 6$ and $w_{3s} = 2$. Which is the starting basic feasible solution of the dual

First Simplex Table

		c_J	1	2	0	0	0	w_B
B	c_B	w_B	Y_1	Y_2	Y_{1s}	Y_{2s}	Y_{3s}	Y_{12}
x_{1s}	0	10	–1	[3]	1	0	0	10/3mini→
x_{2s}	0	6	1	1	0	1	0	6/1
x_{3s}	0	2	1	–1	0	0	1	–ve
			–1	–2	0	0	0	
				↑	↓			

Calculate of $Z_J - c_J$

$\Delta_1 = Z_1 - c_1 = -1$,

$\Delta_2 = -2$

To find incoming vector

Since $\Delta_2 = -2$ is mini of Δ_1 and Δ_2.

So y_2 is incoming vector.

To find out going vector.

Since y_2 is incoming vector so we will find ratio Mini $\left\{\frac{w_B}{y_{i2}}, y_{i2} > 0\right\}$

$$= \text{Mini}\left\{\frac{w_{B1}}{y_{12}}, \frac{w_{B2}}{y_{22}}, \frac{w_{B3}}{y_{32}}\right\}$$

$$= \text{Mini}\left\{\frac{10}{3}, \frac{6}{1}, \frac{2}{-1}\right\}$$

$\frac{10}{3}$ is mini so x_{1s} in out going.

The key element is $y_{12} = 3$

Firstly, we divide the first marked row of the first table by 3 to get new first row of the second table. Calculating second and third row from first row.

Second Table

B	c_B	c_J / w_B	1 / Y_1	2 / Y_2	0 / Y_{1s}	0 / Y_{2s}	0 / Y_{3s}	$\frac{w_B}{y_{11}}$
x_2	2	10/3	–1/3	1	1/3	0	0	–ve
x_{2s}	0	8/3	[4/3]	0	–1/3	1	0	2Mini→
x_{3s}	0	16/3	2/3	0	1/3	0	1	8
			–5/3	0	2/3	0	0	
			↑			↓		

calculate of $Z_J - c_J$

$\Delta_1 = Z_1 - c_1 = -5/3,$

$\Delta_2 = Z_2 - c_2 = 0$

$\Delta_2 = 2_3 - c_3 = 2/3,$

To find incoming vector

Since $\Delta_1 = -5/3$ is mini of Δ_1, Δ_2 Δ_3. So y_1 is incoming vector

To find out going vector.

Since y_1 is incoming vector so we will find ratio Mini $\left\{\frac{w_B}{y_{11}}, y_{11} > 0\right\}$

$$= \text{Mini}\left\{\frac{w_{B1}}{y_{11}}, \frac{w_{B2}}{y_{21}}, \frac{w_{B3}}{y_{32}}\right\}$$

$$= \text{Mini}\left\{-\text{ve}, \frac{8\times 3}{3\times 4}, \frac{16\times 3}{3\times 2}\right\}$$

2 is mini so x_{2s} is out going vector.

The key element is $y_{21} = 4/3$

First we divide the second marked row of the second table by 4/3 to get the new second of the third table.

Calculating first and third row from second row.

Third Simplex Table

B	c_B	c_J / w_B	1 / Y_1	2 / Y_2	0 / Y_{1s}	0 / Y_{2s}	0 / Y_{3s}
x_2	2	4	0	1	1/4	1/4	0
x_1	1	2	1	0	–1/4	3/4	0
x_{3s}	0	2_1	0	0	1/2	–1/2	1
			0	0	1/4	5/4	0

All $Z_J - c_J \geq 0$, so this solution is optimal.

Optimal solution of the dual problem is

$$w_1 = 2$$

$$w_2 = 4$$

$$\text{Max} \quad Z_D = 10$$

Solution of the original problem is

$$x_1 = 1/4$$

$$x_2 = 5/4$$

$$x_3 = 0$$

$$\text{Mini} \quad Z_P = 10.$$

Example 3:

Using the dual, solve the following L.P.P.

$$Max \quad Z_P = 3x_1 - 2x_2$$

$$s.t. \quad x_1 \leq 4$$

$$x_2 \leq 6$$

$$x_1 + x_2 \leq 5$$

$$-x_2 \leq -1$$

$$x_1, x_2 \geq 0.$$

Solution:

The dual of the given problem is given by

$$\text{Mini} \quad Z_D = 4w_1 + 6w_2 + 5w_3 - w_4$$

$$\text{s.t.} \quad w_1 + w_3 \geq 3$$

$w_2 + w_3 - w_4 \geq -2$

and $w_1, w_2, w_3, w_4 \geq 0.$

Now firstly we cell convert the dual problem of minimization to maximization problem by taking the objective function as

Max $Z_D' = -Z_D = -4w_1 - 6w_2 - 5w_3 + w_4$

s.t. $w_1 + w_3 \geq 3$

$-w_2 - w_3 + w_4 \leq 2$

Introducing the surplus, slack variable w_{1s}, w_{2s}, respectively, the constraints of the dual problem reduces to the following equities.

$w_1 + w_3 - w_{1s} = 3$

$-w_2 - w_3 + w_4 + w_{2s} = 2$

$w_1, w_2, w_3, w_4, w_{1s}, w_{2s} \geq 0.$

Since identity matrix I_2 is present in the coefficient matrix. So there is no need to introduce artificial variable in the first constraint.

Taking $w_2 = 0 = w_3 = w_4 = w_{1s}$, we have $w_1 = 3$ and $w_{2s} = 2$. Which is the starting basic feasible solution of the dual.

First Table

B	c_B	c_J / w_B	-4 / Y_1	-6 / Y_2	-5 / Y_3	1 / Y_4	0 / Y_{1s}	0 / Y_{2s}	$\frac{w_B}{Y_{14}}$
x_1	-4	3	1	0	1	0	-1	0	$\frac{3}{0} = \infty$
x_{2s}	0	2	0	-1	-1	[1]	0	1	2 mini→
	$Z_J - c_J$		0	6	1	-1	4	0	
						↑		↓	

Calculate $Z_J - c_J$

$\Delta_1 = Z_1 - c_1 = 0, \Delta_2 = 6, \Delta_3 = 1$

$\Delta_4 = -1, \Delta_5 = 4.$

Since $\Delta_4 = -1$ is mini of $\Delta_1\ \Delta_2\ \Delta_3$ so y_4 is incoming vector.

To find the out going vector since y_4 is incoming vector so we will find ratio Mini $\left\{\frac{w_B}{y_{i4}}, y_{i4} >\right\}$

Mini $\{\infty, 2\}$

2 is mini x_{2s} is out going vector.

The key element is $y_{24} = 1$

Proceeding as before the second simplex table is

Second Simplex Table

		c_J	–4	–6	–5	1	0	0
c_B	w_B	Y_1	Y_2	Y_3	Y_4	Y_{1s}	Y_{2s}	
x_1	–4	3	1	0	1	0	–1	0
x_4	1	2	0	–1	–1	1	0	1
		$Z_J - c_J$	0	5	0	0	4	1

Since all $Z_J - c_J \geq 0$, so this solution of the dual problem is optimal

Optimal solution of the dual problem is

$$w_1 = 3, w_2 = 0$$

$$w_3 = 0, w_4 = 2$$

Mini $\quad Z_D = \text{Max } Z_D' = 10$

Solution of the original problem

$$x_1 = 4$$

$$x_2 = 1$$

Max. $\quad Z_P = 10.$

Example 4:

Find the dual of the following problem and hence solve it.

Mini $\quad Z_P = 6x_1 + 5x_2 + 2x_3$

s.t. $\quad x_1 + 3x_2 = 2x_3 \geq 5$

$$4x_1 - 2x_2 + 3x_3 \geq -1$$

$$2x_1 + 2x_2 + x_3 \geq 2$$

$$x_1, x_2, x_3 \geq 0.$$

Solution:

The dual of this problem is given by

Max. $\quad Z_D = 5w_1 - w_2 + 2w_2$

s.t. $\quad w_1 + 4w_2 + 2w_3 \leq 6$

$3w_1 - 2w_2 + 3w_3 \leq 5$

$2w_1 + 3w_2 + w_3 \leq 2$

$w_1, w_2, w_3 \leq 0.$

Introducing the slack variable w_{1s}, w_{2s}.

DUALITY THEOREMS

We observe that dual of a primal problem is also a linear programming problem therefore we can also construct the dual of a dual problem.

Here we shall prove a number of fundamental theorems to describe the relation between the primal and its dual. The relationship between the primal and dual is extremely useful in the development of mathematical programming.

Theorem 1:

The dual of a dual of the given primal is the primal itself.

Proof:

Suppose the given linear programming problem is

Primal Problem

Max. $Z_P = c_1 x_1 + c_2 x_2 + ... + c_n x_n$

subject to

$$\left.\begin{array}{l} a_{11} x_1 + a_{12} x_2 + ... + a_{1n} x_n \leq b_1 \\ a_{21} x_1 + a_{22} x_2 + ... + a_{2n} x_n \leq b_2 \\ \quad ... \qquad ... \qquad ... \\ \quad ... \qquad ... \qquad ... \\ a_{m1} x_1 + a_{m2} x_2 + ... + a_{mn} x_n \leq b_m, \end{array}\right\} \quad ...(1)$$

$x_1, x_2, ... , m_n \geq 0.$

Dual Problem

The dual of the above primal can be written as

Mini. $Z_D = b_1 w_1 + b_2 w_2 + ... + b_m w_m \geq c_1$

subject to

$a_{11} w_1 + a_{21} w_2 + ... + a_{m1} w_m \geq c_1$

$a_{12} w_1 + a_{22} w_2 + ... + a_{m2} w_m {}^3 c_2$ ...(2)

$$\cdots \qquad \cdots \qquad \cdots$$

$$\cdots \qquad \cdots \qquad \cdots$$

$$a_{1n}\, w_1 + a_{2n}\, w_2 + \ldots + a_{mn}\, w_m\, {}^3\, c_n$$

$$w_1, w_2, \ldots, w_m \geq 0.$$

Now we have to construct the dual of the above dual. First we shall change the above dual to standard maximization form.

$$\text{Max. } (-Z_D) = -b_1\, w_1 - b_2\, w_2 - \ldots - b_m\, w_m$$

subject to

$$\left.\begin{array}{l} -a_{11}\, w_1 - a_{21}\, w_2 - \ldots - a_{m1}\, w_m \leq -c_1 \\ -a_{12}\, w_1 - a_{22}\, w_2 - \ldots - a_{m2}\, w_m \leq -c_2 \\ \quad\cdots \qquad \cdots \qquad \cdots \\ \quad\cdots \qquad \cdots \qquad \cdots \\ -a_{1n}\, w_1 - a_{2n}\, w_2 - \ldots - a_{mn}\, w_m \leq -c_n \end{array}\right\} \quad \ldots(3)$$

$$w_1, w_2, \ldots, w_m \geq 0.$$

Dual of the Dual

Considering the above dual as primal, its dual can be written as

$$\text{Mini. } Z_y = -c_1\, y_1 - c_2\, y_2 - \ldots - c_n\, y_n$$

subject to

$$\left.\begin{array}{l} -a_{11}\, y_1 - a_{12}\, y_2 - \ldots - a_{1n}\, y_n \geq -b_1 \\ -a_{21}\, y_1 - a_{22}\, y_2 - \ldots - a_{2n}\, y_n \geq -b_2 \\ \quad\cdots \qquad \cdots \qquad \cdots \\ \quad\cdots \qquad \cdots \qquad \cdots \\ -a_{m1}\, y_1 - a_{m2}\, y_2 - \ldots - a_{mn}\, y_n \geq -b_m, \end{array}\right\} \quad \ldots(4)$$

$$y_1, y_2, \ldots, y_n \geq 0.$$

Changing the above problem to maximization and multiplying each constraint by – 1, we get

$$\text{Max. } Z_y' = c_1\, y_1 + c_2\, y_2 + \ldots + c_n\, y_n, \; (-Z_y = Z_y' \text{ say})$$

subject to

$$a_{11}\, y_1 + a_{12}\, y_2 + \ldots + a_{1n}\, y_n \leq b_1$$

$$a_{21}\, y_1 + a_{22}\, y_2 + \ldots + a_{2n}\, y_n \leq b_2 \quad \ldots(5)$$

$$\cdots \qquad \cdots \qquad \cdots$$

$$\cdots \qquad \cdots \qquad \cdots$$

$a_{m1} y_1 + a_{m2} y_2 + ... + a_{mn} y_n \le b_m,$

$y_1, y_2, ..., y_n \ge 0.$

Which is identical to the given linear programming problem (primal problem).

Hence the dual of the dual is the primal.

Theorem 2:

If x is any feasible solution to the primal problem Max $Z_P = c\,x$, subject to $A\,x \le b$, $x \ge 0$ and w is any feasible solution to the dual problem

$$\text{Min } Z_D = b'\,w$$

subject to $A'\,w \ge c'$, $w \ge 0$,

then $c\,x \le b'\,w$ i.e., $Z_P \le Z_D$.

Proof:

Consider the primal problem

$$\left.\begin{aligned} &\text{Max. } Z_P = c\,x \\ &\text{subject to } A\,x \le b \\ &\qquad x \ge 0. \end{aligned}\right\} \quad ...(1)$$

Let $x = (x_1, x_2, ..., x_n)$ be any feasible solution to (1).

The dual of the above primal is

$$\left.\begin{aligned} &\text{Min. } Z_D = b'\,w \\ &\text{subject to } A'\,w \ge c',\ w \ge 0. \end{aligned}\right\} \quad ...(2)$$

Let $w = (w_1, w_2, ... w_m)$ be any feasible solution to the dual (2).

Now w (m-component column vector) is the feasible solution of the dual and $A\,x \le b$ are the constraints of the primal (1). If we multiply both sides of $A\,x \le b$ with w', the sign of the inequality remains unchanged.

$$\left.\begin{aligned} &\therefore w'\,(A\,x) \le w'\,b \\ &\Rightarrow (A'\,w)'\,x \le (b'\,w)', \end{aligned}\right\} \quad ...(3)$$

Similarly, x (n-component column vector) is the feasible solution of the primal (1) and $A'w \ge c'$ denotes the constraints of the dual (2) so we can write

$x'\,(A'\,w) \ge x'\,c'$

$\Rightarrow x'\,(w'\,A)' \ge (c\,x)'$

$\Rightarrow [(w'\,A)x]' \ge (c\,x)'$

$\Rightarrow (w'\,A)\,x \ge c\,x$

$\Rightarrow (A' w)' x \geq c x.$ ...(4)

From relations (3) and (4) we have

$c x \geq (A' w)' x \leq (b' w)'$

$\Rightarrow c x \leq (b' w)' \Rightarrow c x \leq b' w$

$\Rightarrow Z_P \leq Z_D.$

Theorem 3:

If $\hat{x}$ is feasible solution to the primal problem Max. $Z_P = c x$ subject to $A x \leq b$, $x \geq 0$ and $\hat{w}$ is a feasible solution to its dual

Min. $Z_D = b' w$ subject to $A' w \geq c'$, $w \geq 0$

such that $c \hat{x} = b' \hat{w}$, then x is the optimal solution of the primal and $\hat{w}$ is the optimal solution of the dual problem.

OR

The necessary and sufficient condition for any linear programming problem and its dual to have optimal solution is that both have feasible solution.

Proof:

Consider the primal problem

$$\left.\begin{array}{l} \text{Max. } Z_P = c x \\ \text{subject to } A x \leq b,\ x \geq 0. \end{array}\right\} \quad ...(1)$$

Let x be any feasible solution to (1).

The dual of the problem (1) is

$$\left.\begin{array}{l} \text{Min. } Z_D = b' w \\ \text{subject to } A' w \geq c',\ w \geq 0 \end{array}\right\} \quad ...(2)$$

Let $\hat{w}$ be any feasible solution to the dual (2).

Proceeding as in theorem (2) above, we have

$c x \leq b' \hat{w}$

$\Rightarrow c x \leq c \hat{x}$, since $c \hat{x} = b' \hat{w}$

$\Rightarrow$value of the objective function of the primal problem at the feasible solution $\hat{x}$ is greater than its value at any other feasible solution $\hat{x}$.

Hence $\hat{x}$ is the optimal solution of the primal problem (for maximization problem).

Again suppose w is any feasible solution of the dual problem (2). x is the given feasible solution of the primal problem (1),

Proceeding as in theorem (2), we have

$c\,x \leq b'\,w$

$\Rightarrow b'\,\hat{w} \leq b'\,w$ since $c\,x = b'\,\hat{w}$

$\Rightarrow$the value of the objective function of the dual problem at the given feasible solution $\hat{w}$ is less than its value at any other feasible solution $\hat{w}$.

Hence $\hat{w}$ is the optimal solution of the dual problem (for minimization problem).

Theorem 4:

(Basic Duality Theorem) : If x_0 is an optimum solution to the primal, then there exists a feasible w_0 to the dual such that

$$c\,x_0 = b'\,w_0,$$

where b′ is the transpose of b.

Proof:

Consider the following primal and dual problems

$$\left.\begin{array}{l} \text{Max. } Z_P = c\,x \\ \text{subject to } A\,x \leq b,\ x \geq 0 \end{array}\right\} \quad \ldots(1)$$

$$\left.\begin{array}{l} \text{and} \quad \text{Min. } Z_D = b'\,w \\ \text{subject to } A'\,w \geq c',\ w \geq 0. \end{array}\right\} \quad \ldots(2)$$

Suppose x_0 is an optimum solution to the primal. Then to solve the primal problem by simplex method, introduce slack variables to each of the constraints.

Now (1) can be written as

$$\text{Max } Z_P = c\,x$$

subject to $A\,x + Ix_s = b$,

where $x_s \in R^m$ represents the vector of slack variables and I is the associated m × n identity matrix.

Suppose $x_0 = (x_B, 0)$ is an optimum solution to the problem where x_B denotes the optimum basic feasible solution.

We know $x_B = B^{-1}\,b$, B is the optimal basis of A.

∴ the optimal primal objective function is

$$Z = c\,x_0 = c_B\,x_B,$$

c_B is the row vector containing the prices of the basic variable.

We know

$$Z_j - c_j = c_B\,Y_j - c_j$$

$$= \begin{cases} c_B B^{-1}\alpha_j - c_j, \forall \alpha_j \in A \\ c_B B^{-1} e_j - 0, \forall e_j \in I \end{cases}$$

Since x_0 is an optimal solution, so we have

$Z_j - c_j \geq \forall j$

$\Rightarrow c_B\ B^{-1}\ a_j \geq c_j,\ c_B\ B^{-1}\ e_j > 0,\ \forall_j$

$\Rightarrow c_B\ B^{-1}\ A \geq c,\ c_B\ B^{-1} \geq 0$ (in matrix form)

$\Rightarrow A'\ B^{-1}\ c_B' \geq c',\ B^{-1}\ c_B \geq 0$

$\Rightarrow A'\ w_0 \geq c';\ w_0 \geq 0$, taking $B^{-1}\ c_B = w_0 \in R^m$

$\Rightarrow w_0$ is a feasible solution of the dual problem. Also corresponding dual objective function is

$b'\ w_0 = b_0'\ w = c_B\ B^{-1}\ b = c_B\ x_B = c\ x_0$

Hence corresponding to a given optimal solution x_0 of the primal there exists a feasible solution w_0 of the dual such that

$c\ x_0 = b'\ w_0$.

Note: Similarly, we can prove the following theorem:

If w_0 is an optimal solution to the dual, then there exists a feasible solution x_0 to the primal such that $c\ x_0 = b'\ w_0$.

Theorem 5:

(Fundamental Duality Theorem). (i) If either the primal or the dual pr*oblem has a finite optimal solution then the other problem so has a finite optimal solution and the optimal values of the objective function in both the problems are the same.*

(ii) If primal (dual) problem has an unbounded optimum solution, the other problem has either no solution at all or an unbounded solution.

Proof:

Consider the primal and the dual problems

Primal problem

$$\left.\begin{aligned} \text{Max. } Z_P &= c\ x \\ \text{subject } A\ x &\leq b,\ x \geq 0 \end{aligned}\right\} \quad \ldots(1)$$

Dual problem.

$$\left.\begin{aligned} \text{Min. } Z_D &= b'\ w \\ \text{subject to } A'\ w &\geq c' \end{aligned}\right\} \quad \ldots(2)$$

$w \geq 0$

To prove the theorem we shall construct an optimal solution to the dual form a given optimal solution of the primal.

Let us assume that the primal has a finite optimal feasible solution x_B.

To solve the primal (1) by simplex method, introduce slack variables to each of the constraints.

Now (1) can be written as

$$\left.\begin{aligned} &\text{Max. } Z_P = c\,x \\ &\text{subject } A\,x + Ix_s = b \\ &x \geq 0,\ x_s \geq 0, \end{aligned}\right\} \qquad \text{...(3)}$$

where $x_s \in R^m$ represents the vector of slack variables and I is the associated m × n identity matrix. B is the basis matrix and suppose c_B is the m-component row vector containing the prices of the basic variables.

Now x_B is the optimal solution to the primal so we have

$c_j - Z_j \leq 0, \forall_j$.

We know $Z_j = c_B Y_j = c_B B^{-1} \alpha_j$

$c_j - c_B B^{-1} \alpha_j \leq 0, \forall a_j$

or $\quad c_B B^{-1} \alpha_j \geq c_j$ ...(4)

$\Rightarrow c_B B^{-1} (\alpha_1, \alpha_2, \ldots, \alpha_n) \geq (c_1, c_2, \ldots, c_n)$

$\Rightarrow c_B B^{-1} A \geq c$ ...(5)

Taking $(\hat{w})' = c_B B^{-1}$, where $\hat{w} = (w_1, w_2, \ldots, w_m)$.

Then from (5), we have

$(w)' A \geq c \Rightarrow [(w)' A]' \geq c'$

$\Rightarrow A' (\hat{w}) \geq c'$

$\Rightarrow \hat{w}$ is the solution of the dual (2).

Again considering the relation (4) with a_j corresponding to slack variables, we have

$c_B B^{-1} (e_j) \geq 0, j = 1, 2, \ldots, m; c_j = 0.$

$\Rightarrow (\hat{w})' e_j \geq 0 \Rightarrow (\hat{w})' \geq 0$

$\Rightarrow \hat{w} \geq 0$

$\Rightarrow$w is the feasible solution of the dual (2).

Now it remains to show that w is an optimal solution to the dual (2).

We have

$Z_D = b' \hat{w} = [(\hat{w})' b]' = (\hat{w})' b$

$= (c_B B^{-1}) b = c_B (B^{-1} b)$

$= c_B x_B = Z_P.$

Since w and x_B are the feasible solution of the dual (2) and the primal (1) respectively and

$Z_D = Z_P$

therefore $\hat{w}$ is the optimal solution of the dual (2). (See the result of theorem 3)

(ii) We shall prove it by contradiction. Suppose, when the primal problem has an unbounded solution, the dual problem has a finite optimal solution.

We have proved in theorem 1 that dual of the dual is the original primal. Also part (i) of this theorem we have proved that if a primal has a finite optimal solution, the dual also has finite optimal solution.

Thus if we consider the dual as the primal, then its (which is the original primal) must have a finite optimal solution, which contradicts the hypothesis. Hence the dual has no finite optimal solution. If the primal problem has an unbounded solution, either the dual has no solution or an unbounded solution.

Similarly, we can prove that if the dual has an unbounded solution, the primal has no solution or an unbounded solution.

Theorem 6:

If any of the constraints in the primal is a perfect equality, the corresponding dual variable is unrestricted in sign.

Proof:

Suppose in the given primal the kth constraint is an equality. Writing the primal in the standard primal form, we have

Max. $Z = c_1 x_1 + c_2 x_2 + \ldots + c_n x_n$

subject to

$a_{11} x_1 + a_{12} x_2 + \ldots + a_{1n} x_n \leq b_1$

$a_{21} x_1 + a_{22} x_2 + \ldots + a_{2n} x_n \leq b_2$

$\ldots \qquad \ldots \qquad \ldots$

$\ldots \qquad \ldots \qquad \ldots$

$a_{k1} x_1 + a_{k2} x_2 + \ldots + a_{kn} x_n \leq b_k$

$$- a_{k1} x_1 - a_{k2} x_2 - \ldots - a_{kn} x_n \leq - b_k$$

$$\ldots \quad \ldots \quad \ldots$$

$$\ldots \quad \ldots \quad \ldots$$

$$a_{m1} x_1 + a_{m2} x_2 + \ldots + a_{mn} x_n \leq b_m,$$

$$x_1, x_2, \ldots, x_n \geq 0.$$

The dual of the above primal is

$$\text{Min } Z_D = b_1 w_1 + b_2 w_2 + \ldots + b_k (w_k' \; w_k'') + \ldots + b_m w_m$$

subject to

$$a_{11} w_1 + a_{21} w_2 + \ldots + a_{k1} (w_k' - w_k'') + \ldots + a_{m1} w_m \geq c_1$$

$$a_{12} w_1 + a_{22} w_2 + \ldots + a_{k2} (w_k' - w_k'') + \ldots + a_{m2} w_m \geq c_2$$

$$\ldots \quad \ldots \quad \ldots$$

$$\ldots \quad \ldots \quad \ldots$$

$$a_{1n} w_1 + a_{2n} w_2 + \ldots + a_{kn} (w_k' - w_k'') + \ldots + a_{mn} w_m \geq c_n$$

$$w_1, w_2, \ldots, w_k', w_k'', \ldots, w_m \geq 0.$$

Substituting $w_k = w_k' - w_k''$ in the above dual, we get

$$\text{Min. } Z_D = b_1 w_1 + b_2 w_2 + \ldots + b_k w_k + \ldots + b_m w_m$$

subject to

$$a_{11} w_1 + a_{21} w_2 + \ldots + a_{ki} w_k + \ldots + a_{m1} w_m \geq c_1$$

$$a_{12} w_1 + a_{22} w_2 + \ldots + a_{k2} w_k + \ldots + a_{m2} w_m \geq c_2$$

$$\ldots \quad \ldots \quad \ldots$$

$$\ldots \quad \ldots \quad \ldots$$

$$a_{1n} w_1 + a_{2n} w_2 + \ldots + a_{kn} w_k + \ldots + a_{mn} w_m \geq c_n$$

$$w_1, w_2, \ldots, w_k - 1, w_k + 1, \ldots, w_m \geq 0.$$

wk is unrestricted in sign because

$$w_k > 0 \text{ if } w_k' > w_k'' \text{ and } w_k < 0 \text{ if } w_k' < w_k''.$$

Theorem 7:

If any variable of the primal is unrestricted in sign, the corresponding constraint in the dual will be a strict equality.

Proof:

Consider the primal in which the kth variable is unrestricted in sign.

$$\text{Max. } Z = c_1 x_1 + c_2 x_2 + .. + c_k x_k + \ldots + c_n x_n$$

subject to

$$a_{11} x_1 + a_{12} x_2 + \dots + a_{1k} x_k + \dots + a_{1n} x_n \le b_1$$

$$a_{21} x_1 + a_{22} x_2 + \dots + a_{2k} x_k + \dots + a_{2n} x_n \le b_2$$

$$\dots \quad \dots \quad \dots$$

$$\dots \quad \dots \quad \dots$$

$$a_{m1} x_1 + a_{m2} x_2 + \dots + a_{mk} x_k + \dots + a_{mn} x_n \le b_m$$

$$x_1, x_2, \dots, x_k - 1, x_{k+1} \dots, x_n \ge 0,$$

x_k is unrestricted in sign.

Substituting $x_k = x_k' - x_k''$, $x_k' \ge 0$ in the given primal, it changes to

$$\text{Max. } Z = c_1 x_1 + c_2 x_2 + \dots + c_k (x_k' - x_k'') + .. + c_n x_n$$

subject to

$$a_{11} x_1 + a_{12} x_2 + \dots + a_{1k} (x_k' - x_k'') + \dots + a_{1n} x_n \le b_1$$

$$a_{21} x_1 + a_{22} x_2 + \dots + a_{2k} (x_k' - x_k'') + \dots + a_{2n} x_n \le b_2$$

$$\dots \quad \dots \quad \dots$$

$$\dots \quad \dots \quad \dots$$

$$a_{m1} x_1 + a_{m2} x_2 + \dots + a_{mk} (x_k' - x_k'') + \dots + a_{mn} x_n \le b_m,$$

$$x_1, x_2, \dots, x_{k-1}, x_k', x_k'', x_{k+1}, \dots, x_n \ge 0.$$

Writing the dual of the above primal, we have

$$\text{Min. } Z_D = b_1 w_1 + b_2 w_2 + .. + b_m w_m,$$

subject to

$$a_{11} w_1 + a_{21} w_2 + \dots + a_{m1} w_m \ge c_1$$

$$a_{12} w_1 + a_{22} w_2 + \dots + a_{m2} w_m \ge c_2$$

$$\dots \quad \dots \quad \dots$$

$$\dots \quad \dots \quad \dots$$

$$a_{1k} w_1 + a_{2k} w_2 + \dots + a_{mk} w_m \ge c_k$$

$$- a_{1k} w_1 - a_{2k} w_2 - \dots - a_{mk} w_m \ge - c_k$$

$$\dots \quad \dots \quad \dots$$

$$\dots \quad \dots \quad \dots$$

$$a_{1n} w_1 + a_{2n} w_2 + \dots + a_{mn} w_m \ge c_n,$$

$$w_1, w_2, \dots, w_m \ge 0.$$

The two constraints

$$a_{1k} w_1 + a_{2k} w_2 + \dots + a_{mk} w_m \ge c_k$$

and $\quad -a_{1k}w_1 - a_{2k}w_2 - \ldots - a_{mk}w_m \geq -c_k$

are equivalent to the single equation

$$a_{1k}w_1 + a_{2k}w_2 + \ldots + a_{mk}w_m = c_k.$$

Hence if the kth variable of the primal is unrestricted, the kth constraint in the dual is an equality.

EXISTENCE THEOREMS

1. *There exist a bounded (finite) optimum solution to a linear programming problem if and only if there exists a feasible solution to both primal and its dual.*
2. *If there does not exist feasible solution to the dual (primal) but there exists at least one to the primal (dual), then there dies not exist any finite optimum solution to the primal (dual).*
3. *If there does not exist any finite optimum solution to the primal (dual) then there does not exist any feasible solution to the dual (primal).*

Proof:

1. Suppose Max. $Z = c\,x$

   subject to $A\,x \leq b,\ x \geq 0$

   and $\quad$ Min. $Z_D = b'\,w$

   subject to $A'\,w \geq c',\ w \geq 0$

   are the primal and dual problems respectively.

   Again suppose there exists an optimum feasible solution to the primal problem. Then by fundamental theorem of duality the dual problem has atleast one feasible solution. Conversely, let us assume that both primal and the dual possess feasible solution x and w respectively. Then $c\hat{x}$ and $b'\hat{w}$ are both finite and $c\,\hat{x} \leq b'\,\hat{w}$ *i.e.*, $b'\,\hat{w}$ acts as an upper bound on $c\,\hat{x}$ although not necessarily the least upper bound. Hence the primal must have finite optimum solution.

2. Consider the primal problem

   $$\text{Max. } Z = c\,x,$$

   subject to $A\,x \leq b,\ x \geq 0$.

   The corresponding dual problem is

   $$\text{Min. } Z_D = b'\,w$$

   subject to $A'\,w \geq c',\ w \geq 0$.

Suppose there does not exist any feasible solution to the dual there does one to the primal, say $\hat{x}$. Then c $\hat{x}$ is the value of the primal objective function.

If we suppose that $\hat{x}$ is an optimum solution to the primal then by fundamental theorem of duality there must exist a feasible solution to the dual which contradicts the hypothesis.

Therefore no feasible solution to the primal can be optimal.

Similarly, we can start with the primal.

3. Suppose the primal problem does not possess any finite optimum solution. Without loss of generality, it can be assumed that there exists a feasible solution to the primal. Now, if we assume that the dual problem possesses a feasible solution then by existence theorem 1 (above) the primal problem must posses a finite optimum solution. This constricted the hypothesis, hence the dual problem does not possess a feasible solution.

COMPLEMENTARY SLACKNESS THEOREM

For the optimal feasible solution of the primal and the dual systems

(i) *If the inequality occurs in the ith relation of either system (the corresponding slack or surplus variable if positive), then the ith variable of its dual is zero.*

(ii) *If the ith variable if positive in either system, the ith relation of its dual holds as a strict equality (i.e., the corresponding slack or surplus variable $w_{m \cdot j}$ = 0.)*

Proof:

The objective functions of the primal and dual problems in explicit form can be written as

$$\text{Max. } Z_P = c_1 x_1 + c_2 x_2 + \ldots + c_n x_n \text{ (primal)} \qquad \ldots(1)$$

$$\text{Min. } Z_D = b_1 w_1 + b_2 w_2 + \ldots + b_m w_m \text{ (dual)} \qquad \ldots(2)$$

After introducing the non-negative slack variables $x_{n+1}, x_{n+2}, \ldots, x_{n+m}$ the primal constraint equations can be written as

$$\left.\begin{array}{l} a_{11} x_1 + a_{12} x_2 + \ldots + a_{1n} x_n + x_{n+1} = b_1 \\ a_{21} x_1 + a_{22} x_2 + \ldots + a_{2n} x_n + x_{n+2} = b_2 \\ \ldots \qquad \ldots \qquad \ldots \\ \ldots \qquad \ldots \qquad \ldots \\ a_{m1} x_1 + a_{m2} x_2 + \ldots + a_{mn} x_n + x_n + m = b_m \end{array}\right\} \qquad \ldots(3)$$

$$x_1, x_2, \ldots, x_{n+m} \geq 0.$$

Again introducing the no-negative surplus variables $w_{m+1}, w_{m+2}, \ldots, w_{m+n}$ the dual constraints equations can be written as

$$\left.\begin{aligned}
a_{11} w_1 + a_{21} w_2 + \ldots + a_{m1} w_m - w_{m+1} &= c_1 \\
a_{12} w_1 + a_{22} w_2 + \ldots + a_{m2} w_m - w_{m+2} &= c_2 \\
\ldots \quad \ldots \quad \ldots & \\
\ldots \quad \ldots \quad \ldots & \\
a_{1n} w_1 + a_{2n} w_2 + a_{mn} w_m - w_{m+n} &= c_n \\
w_1, w_2, \ldots, w_m, w_{m+1}, \ldots, w_{m+n} &\geq 0.
\end{aligned}\right\} \quad \ldots(4)$$

Now, multiplying the equations (3) by $w_1, w_2, \ldots, w_m$ respectively and then adding the resulting equations, we have

$$x_1 \sum_{i=1}^{m} a_{i1} w_i + x_2 \sum_{i=1}^{m} a_{i2} w_i + \ldots + x_n \sum_{i=1}^{m} a_{in} w_i$$

$$+ x_{n+1} w_1 + x_{n+2} w_2 + \ldots + x_{n+m} w_m = \sum_{i=1}^{m} b_i w_i \quad \ldots(5)$$

subtracting (5) from (1), we have

$$\left(c_1 - \sum_{i=1}^{m} a_{i1} w_i\right) x_1 + \left(c_2 - \sum_{i=1}^{m} a_{i2} w_i\right) x_2 + \ldots + \left(c_n - \sum_{i=1}^{m} a_{in} w_i\right) x_n$$

$$- w_1 x_{n+1} - w_2 x_{n+2} - \ldots - w_m x_{n+m} = \left(Z_P - \sum_{i=1}^{m} b_i w_i\right)$$

$$= Z_P - Z_D \quad \ldots(6)$$

From (4), we have

$$- w_{m+j} = c_j - (a_{1j} w_1 + a_{2j} w_2 + \ldots + a_{mj} w_m), \; j = 1, 2, \ldots, n$$

$$\text{or} - w_{m+j} = c_j - \sum_{i=1}^{m} a_{ij} w_i \text{ for all } j = 1, 2, \ldots, n. \quad \ldots(7)$$

Using (7) in (6), we have

$$- (w_{m+1} x_1 + w_{m+2} x_2 + \ldots + w_{m+n} x_n)$$

$$- (w_1 x_{n+1} + w_2 x_{n+2} + \ldots + w_m x_{n+m}) = Z_P - Z_D \quad \ldots(8)$$

Now if $x^* = (x_1^*, x_2^*, \ldots, x_n^*)$ and $w^* = (w_1^*, w_2^*, \ldots, w_m^*)$ be the optimal solution to the primal and the dual problems respectively, then by Duality theorem

$$Z_P^* = Z_D^*.$$

Thus for the optimal solution x* and w* of the primal and dual problems the corresponding slack and surplus variables are

$$x^*_{n+1} \geq 0,\ i = 1, 2, \ldots, m \text{ and } w^*_{m+j} \geq 0, = 1, 2, \ldots, n.$$

Form equation (8), we have

$$(w^*_{m+1}\, x_1^* + w^*_{m+2}\, x_2^* + \ldots + w^*_{m+n}\, x_n^*)$$

$$+ (w_1^*\, x^*_{n+1} + w_2^*\, x^*_{n+2} + \ldots + w_m^*\, x^*_{n+m}) = 0 \qquad \ldots(9)$$

Since all the variables in (9) are non-negative so their product terms in (9) are also non-negative. For the validity of relation (9) each term must be individually equal to zero

i.e., $\quad w^*_{m+j}\, x_j^* = 0\ (\forall_j = 1, 2, \ldots, n) \qquad \ldots(10)$

and $\quad w_i^*\, x^*_{n+i} = 0\ (\forall_i = 1, 2, \ldots, m) \qquad \ldots(11)$

(i)Form (11), if $x^*_{n+1} > 0$ then we must have $w_i^* = 0$ *i.e.*, if the slack variable in the ith relation of the primal is positive then ith variable of the dual is zero.

Again from (10), if $w^*_{m+j} > 0$ then we must have $x_j^* = 0$ *i.e.*, if the surplus variable in the ith relation of the dual is positive then jth variable of the primal (dual of the dual) is zero.

This proves the part (i) of the theorem.

(ii) Also from (10), if $x_j^* > 0$, then we must have $w^*_{m+j} = 0$ *i.e.*, if the ith variable in the primal is positive then the jth relation in the dual is strictly an equality (as $w^*_{m+j} = 0$).

Again from (11), if $w_i^* > 0$, then $x^*_{n+j} = 0$

i.e., if the ith variable in the dual is positive then the jth relation in the primal (dual of the dual) is strictly an equality (as $x^*_{n+i} = 0$).

This proves the part (ii) of the theorem.

Alternative Statement

A necessary and sufficient condition for any pair of feasible solutions to the primal and dual to be optimal is that

$w_i x_{n+i} = 0$, $i = 1, 2, \ldots, m$ where x_{n+i} is the slack variable in the primal

and $\quad x_j w_{m+j} = 0$, $j = 1, 2, \ldots; n$, where w_{m+j} is the surplus variables for the dual.

Primal and Dual Correspondence

With the help of the duality theorems developed in 7 and 8, we note the following correspondence between the primal and the dual problems.

Primal	*Dual*
1. Objective function Max. Z_P.	Objective function Min. Z_D.
2. Requirement vector.	Price vector.
3. Coefficient matrix A	Transpose of the coefficient matrix, A' or A^T.
4. Constraint with ≤ sign.	Constraints with ≥ sign.
5. Relation.	Variable.
6. i-th inequality.	i-th variable $w_i \geq 0$.
7. i-th constraint an equality.	i-th variable unrestricted in sign .
8. Variable.	Relation.
9. i-th variable $x_i > 0$.	i-th relation a strict inequality.
10. i-th variable x_i unrestricted in sign.	i-th constraint a strict equality.
11. i-th slack variable positive.	i-th variable zero.
12. i-th variable zero.	i-th surplus variable positive.
13. Finite optimal solution.	Finite optimal solution with equal optimal value of objective function.
14. Unbounded solution	No solution or an unbounded solution.

Rules for Obtaining the Solution to the Dual from the Final Simplex Table of the Primal and Vice-Versa

From the final simplex table of the primal problem we can also read the optimal solution of the dual and vice-versa. For this we observe the following rules:

1. *The optimal value of the primal objective function is equal to the optimal value of the dual objective function.*

   i.e., $\quad Max.\ Z_P = Min.\ Z_D.$

   This has been proved in Duality Theorem.

2. *The Δ_j's ($\Delta_j = c_j - Z_j$) with sign changed for the lack (or surplus) variables in the final simplex table for the primal are the values of the corresponding optimal dual variables in the final simplex table for the dual.*

Proof:

In the final simplex table (for primal) corresponding to slack and surplus variables

$$-\Delta_j = -(c_j - Z_j) = -(0 - Z_j)$$

$= c_B Y_j = c_B B^{-1} \alpha_j$

where B is the optimal basis and c_B, the corresponding price vector.

But, for i-th slack variable, a_j becomes a unit vector e_i with 1 at the i-th place.

$\therefore -\Delta_j = c_B B^{-1} e_i = (w_1, w_2, \ldots, w_m) e_i = w_i$

where w_j is i-th dual variable.

3. If either problem has unbounded solution, then the other will have no feasible solution.

Note: It is always advantageous to apply the simplex method to the problem having lesser number of constraints. Therefore, first we shall solve the problem (primal or dual) with lesser number of constraints by simplex method and then read the solution of the other from the final simplex table according to the rules described above.

SOME USEFUL ASPECTS OF DUALITY

In some cases it is easier to solve linear programming problems through their duals. If the number of original variables in the primal problem is considerably less than the number of slack and surplus variables, we must opt to solve the problem through its dual.

Duality plays an important role not only in linear programming but also in physics, economics, engineering etc.

In physics, we use it in parallel circuit and series circuit theory.

In Economics it is used in the formulation of input and output systems.

In game theory, we use it to find the optimal strategies of the player B when he minimizes his losses. Then, using duality, we can change the player A's problem into player B's problem and vice-versa.

Example 1:

Write the dual of following problem and solve it.

$$\text{Max. } Z = 4x_1 + 2x_2$$

subject to $\quad -x_1 - x_2 \le -3$

$$-x_1 + x_2 \le -2,\ x_1,\ x_2 \ge 0.$$

Hence or otherwise write down the solution of the primal.

Solution:

The given problem is in standard primal form.

∴ the dual to the given primal is

Min. $Z_D = -3w_1 - 2w_2$

subject to $-w_1 - w_2 \geq 4$

$-w_1 + w_2 \geq 2$

$w_1, w_2 \geq 0.$

Changing the objective function to maximization and introducing surplus variables $w_3 \geq 0$, $w_4 \geq 0$ and artificial variables $w_5, w_6 \geq 0$ the above problem reduces to

Max. $Z_D = 3w_1 + 2w_2 + 0w_3 + 0w_4 - Mw_5 - Mw_6$

subject to $-w_1 - w_2 - w_3 + w_5 = 4$

$-w_1 + w_2 - w_4 + w_6 = 2$

The starting B.F.S. is $w_1 = 0 = w_2 = w_3 = w_3$, $w_5 = 4$, $w_6 = 2$.

Now applying the simplex method to obtain the optimal solution, we have

		c_j	3	2	0	0	–M	–M	Min. ratio
B	c_B	w_B	W_1	W_2	W_3	W_4	W_5	W_6	w_B/W_2
W_5	–M	4	–1	–1	–1	0	1	0	–ive
W_6	–M	2	–1	[1]	0	–1	0	1	2
$Z_D = -6_M$		Δ_j	3 – 2M	2	–M	–M	0	0	
W_5	–M	6	–2	0	–1	–1	1	1	
W_2	2	2	–1	1	0	–1	0	1	
$Z_D = 4 - 6M$		Δ_j	5 – 2M	0	–M	2 – M	0	–2	

In the last simplex table no $\Delta_j > 0$ and a non-zero artificial variable appears in the basis therefore the dual problem does not possess any optimum basic feasible solution. Consequently, the given problem does not possess any finite optimal solution.

And w_{3s}. The dual problem can be written as

Max. $Z_D = 5w_1 - w_2 + 2w_2$

s.t. $w_1 + 4w_2 + 2w_3 + w_{1s} = 6$

$3w_1 - 2w_2 + 3w_3 + w_{2s} = 5$

$2w_1 + 3w_2 + w_3 + w_{3s} = 2$

$w_1, w_2 - w_{3s} \geq 0.$

Taking $w_1 = 0$ $w_2 = w_3$, we have $w_{1s} = 6$, $w_{2s} = 5$, $w_{3s} = 2$ which is the starting basic feasible solution of the dual.

First Simplex Table

B	c_B	c_J / w_B	5 Y_1	−1 Y_2	2 Y_3	0 Y_{1s}	0 Y_{2s}	0 Y_{3s}	$\frac{w_B}{Y_{11}}$
x_{1s}	0	6	1	4	2	1	0	0	6/1
x_{2s}	0	5	3	−2	2	0	1	0	5/3
x_{3s}	0	2	[2]	3	1	0	0	1	1 mini
			−5	1	−2	0	0	0	
			↑					↓	

Calculate of $Z_J - c_J$

$\Delta_1 = -5$, $\Delta_2 = 1$,

$\Delta_3 = -2$

Since $\Delta_1 = -5$ is mini of Δ_1, Δ_2, Δ_3 so y_1 is incoming vector

To find out going vector

Since y_1 is incoming vector so we will find ratio Mini. $\left\{\frac{w_B}{y_{i1}}, y_{i1} > 0\right\}$

Mini. $\left\{\frac{6}{1}, \frac{5}{3}, \frac{2}{2}\right\}$

1 is mini so y_{3s} is out going vector.

The key element is $y_{31} = 2$

Proceeding as before the second simplex table

Second Table

B	c_B	c_J / w_B	5 Y_1	−1 y_2	2 Y_3	0 Y_{1s}	0 Y_{2s}	0 Y_{3s}
x_{1s}	0	5	0	5/2	3/2	1	0	−1/2
x_{2s}	0	2	0	13/2	1/2	0	1	−3/2
x_1	5	1	1	3/2	1/2	0	0	1/2
		$Z_J - C_J$	0	17/2	5/2	0	0	5/2

Since all $Z_J - c_J \geq 0$. Therefore solution is optimal.

i.e., optimal solution of the dual problems

$w_1 = 1,$

$w_2 = 0,$

$w_3 = 0$

Max. $Z_D = 5$

solution of the primal problem is

$x_1 = 0, x_2 = 0$

$x_3 = 5/2$

Mini. $Z_P = 5.$

Example 2:

Formulate the following L.P.P., into, its dual problem and solve it

Max $Z = x_1 - x_2$

s.t., $2x_1 + x_2 \geq 2$

$-x_1 - x_2 \geq 1$

and $x_1, x_2 \leq 0.$

Solution:

The given L.P.P. is maximizations so all sign should be less than

$-2x_1 - x_2 \leq -2$

$x_2 + x_2 \leq -1$

The dual of this problems is given by

Mini $Z_D = -2w_1 - w_2$

s.t., $-2w_1 + w_2 \geq 1$

$-w_1 + w_2 \geq -1$

Now firstly we will convert the dual problem of minimization to maximizations problem by taking the objective function as

Max $Z_D' = -Z_D = 2w_1 + w_2$

s.t., $-2w_1 + w_2 \geq 1$

$w_1 - w_2 \leq 1$

Introducing the surplus, slack and artificial variable w_{1s}, w_{2s}, w_{1a} respectively. The dual problem can be written as

Max $Z_D' = -Z_D = 2w_1 + w_2$

s.t., $-2w_1 + w_2 - w_{1s} + w_{1a} = 1$

$w_2 - w_2 + w_{2s} = 1$

Taking $w_1 = 0 = w_2 = w_{1s}$, we have $w_{1a} = 1$ $w_{2s} = 1$, which is the starting B.F.S.

First Table

B	c_B	c_J / w_B	2 / Y_1	1 / Y_2	0 / Y_{1s}	0 / Y_{2s}	–M / Y_{1a}	w_B / Y_{i2}
x_{1a}	–M	1	–2	[1]	–1	0	1	1
x_{2s}	0	1	1	–1	0	1	0	–ive
$Z_J–c_J$			$2_M–2$ ↑	–M–1	M	0	0 ↓	

$\Delta_1 = 2M - 2,$

$\Delta_2 = -M - 1, \ \Delta_3 = M$

Since $\Delta_2 = -M - 1$ is mini of Δ_1, Δ_3. So y_2 is incoming vector

To find out doing vector.

By mini ratio rule we find that y_{1a} is out going vector.

The key element is 1

Proceeding as before second simplex table

Second Table

B	c_B	c_J / w_B	2 / Y_1	1 / Y_2	0 / Y_{1s}	0 / Y_{2s}	–M / Y_{1a}	w_B / y_{i3}
x_2	1	1	–2	1	–1	0	1	–ive
x_{2s}	0	2	–1	0	(1)	1	1	2→
		0	0	–1 ↑	0 ↓	1 + M		

Proceeding as usual the third table as follows:

Third Table

B	c_B	c_J / w_B	2 / Y_1	1 / Y_2	0 / Y_{1s}	0 / Y_{2s}	–M / Y_{1a}	w_B / y_{i1}
x_2	1	3	–3	1	0	1	2	–ive
x_{1s}	0	2	–1	0	1	1	1	–ive
$Z_J–c_J$			–5 ↑	0	0	1	2+M	

Because all value of a are (–ve) therefore solution is no feasible solution.

Example 3:

Give the dual of the following problem and solve

$$\text{Mini. } Z_p = x_1 + x_2$$
$$2x_1 + x_2 \geq 4$$
$$x_1 + 7x_2 \geq 7$$
$$x_1, x_2 \geq 0.$$

Solution:

The dual of the given problem is given by

$$\text{Max. } Z_D = 4w_1 + 7w_2$$
$$\text{s.t. } 2w_1 + w_2 \leq 1$$
$$w_1 + 7w_2 \leq 1$$
$$w_1, w_2 \geq 0$$

Introducing the slack variable w_{1s} and w_{2s} respectively. The dual problem can be written as

$$\text{Max. } Z_D = 4w_1 + 7w_2$$
$$\text{s.t. } 2w_1 + w_2 + w_{1s} = 1$$
$$w_1 + 7w_2 + w_{2s} = 1$$
$$w_1, w_2, w_{1s}, w_{2s} \geq 0$$

Taking $w_1 = 0 = w_2$, we have $w_{1s} = 1$, $w_{2s} = 1$, which is the starting basic feasible solution of the dual.

First Table

		c_J	4	7	0	0	$\frac{w_B}{y_{i2}}$
B	c_B	w_B	Y_1	Y_2	Y_{1s}	Y_{2s}	
x_{1s}	0	4	2	1	1	0	4
x_{2s}	0	7	1	[7]	0	1	1 mini
	$Z_J - c_J$		–4	–7	0	0	

$\Delta_1 = -4$, $\Delta_2 = -7$

Since $\Delta_2 = -7$ is mini of Δ_1 and Δ_2 so y_2 is incoming vector and by minimum ratio rule outgoing vector is y_{2s}.

The key element is 7.

Proceeding as before the second simplex table.

Second Table

B	c_B	c_J / w_B	4 / Y_1	7 / y_2	0 / Y_{1s}	0 / Y_{2s}	w_B / Y_{i1}
x_{1s}	0	3	13/7	0	1	–1/7	21/13 mini
x_2	7	1	1/7	1	0	1/7	7
		$Z_J - c_J$	–3 ↑	0	0 ↓	1	

Since $\Delta_1 = -3$ is mini so y_1 is incoming vector and by mini rule method outgoing vector is y_{1s}.

The key element is $y_{11} = 13/7$

Proceeding as before the third simplex table.

Third Table

B	c_B	c_J / w_B	4 / Y_1	7 / Y_2	0 / Y_{1s}	0 / Y_{2s}
x_1	4	21/13	1	0	7/13	–1/13
x_2	7	10/13	0	1	–1/13	1/13
		$Z_J - c_J$	0	0	$\frac{21}{13}$	10/13

Since all $Z_J - c_J \geq 0$, so this solution of the dual problem is optimal solution of the dual is

$$w_1 = 21/13,\ w_2 = 10/13$$

$$\text{Max } Z_D = \frac{31}{13}$$

solution of the primal problem is

$$x_1 = \frac{21}{13},\ x_2 = 10/13$$

Mini. $Z_P = 31/13$.

Example 4:

Write the dual of L.P.P. and solve it.

$$\text{Mini } \quad Z = 8x_1 + 2x_2 + 4x_3$$

$$\text{s.t.} \quad x_1 - 4x_2 - 2x_3 \geq 2$$

$$x_1 + x_2 - 3x_3 \geq -1$$

$-3x_1 - x_2 + x_3 \geq 1$

$x_1, x_2, x_3 \geq 0.$

Solution:

The dual of this problem is given by

Max. $Z_D = 2w_1 - w_2 + w_3$

s.t. $w_1 + w_2 - 3w_3 \leq 8$

$-4w_1 + w_2 - w_3 \leq 2$

$-2w_1 - 3w_2 + w_3 \leq +4$

$w_1, w_2, w_3 \geq 0$

Introducing the slack variable w_{1s}, w_{2s} and w_{3s} respectively. The dual problem can be written as

Max $Z_D = 2w_1 - w_2 + w_1$

s.t. $w_1 + w_2 - 3w_3 + w_{1s} = 8$

$-4w_1 + w_2 - w_3 + w_{2s} = 2$

$-2w_1 - 3w_2 + w_3 + w_{3s} = +4$

$w_1, w_2, w_3, w_{1s}, w_{2s}, w_{3s} \geq 0$

Taking $w_1 = 0 = w_2 = w_3$ we have $w_{1s} = 8$, $w_{2s} = 2$ $w_{3s} = 4$, which is the starting basic feasible solution of the dual.

First Table

		c_j	2	–1	1	0	0	0	w_B
B	c_B	w_B	Y_1	Y_2	Y_3	Y_{1s}	Y_{2s}	Y_{3s}	y_{i1}
x_{1s}	0	8	[1]	1	–3	1	0	0	8
x_{2s}	0	2	–4	1	–1	0	1	0	–ive
x_{3s}	0	4	–2	–3	1	0	0	1	–ive
			–2	1	–1	0	0	0	
			↑			↓			

Since $\Delta_1 = -2$ is mini of $\Delta_1, \Delta_2, \Delta_3$. So y_1 is incoming vector and by mini ratio rule out going vector is y_{1s}.

The key element is $y_{11} = 1$

Proceeding as before the second simplex table

Second Table

B	c_B	c_j / w_B	2 / Y_1	–1 / Y_2	1 / Y_3	0 / Y_{1s}	0 / Y_{2s}	0 / Y_{3s}	$\frac{w_B}{y_{i3}}$
x_1	2	8	[1]	1	–3	1	0	0	–ive
x_{2s}	0	34	0	5	–13	4	1	0	–ive
x_{3s}	0	20	0	–1	–5	2	0	1	–ive
		•	0	3	–7 ↑	2	0	0	

all $\frac{w_B}{y_{i3}}$ is negative so solution is no feasible solution or unbounded solution.

Example 5:

Write the dual of L.P.P. and solve it.

$$\text{Mini } Z_P = 3x_1 - 2x_2 + 4x_3$$

$$\text{s.t. } 3x_1 + 5x_2 + 4x_3 \geq 7$$

$$6x_1 + x_2 + 3x_3 \geq 4$$

$$7x_1 - 2x_2 - x_3 \leq 10$$

$$x_1 - 2x_2 + 5x_3 \geq 3$$

$$4x_1 + 7x_2 - 2x_3 \geq 2$$

$$x_1, x_2, x_3 \geq 0.$$

Solution:

In standard form the problem can be written as

$$\text{Mini } Z_P = 3x_1 + 2x_2 + 4x_2$$

$$\text{s.t. } 3x_1 + 5x_2 + 4x_3 \geq 7$$

$$6x_1 + x_2 + 3x_3 \geq 4$$

$$-x_1 + 2x_2 + x_3 \geq -10$$

$$x_1 - 2x_2 + 5x_3 \geq 3$$

$$4x_1 + 7x_2 - 2x_3 \geq 2$$

$$x_1, x_2, x_3 \geq 0$$

The dual of this problem can be written as

$$\text{Max} \quad Z_D = 7w_1 + 4w_2 - 10w_3 + 3w_4 + 2w_5$$

$$\text{s.t.} \quad 3w_1 + 6w_2 - 7w_3 + w_4 = 4w_5 \leq 3$$

$$5w_1 + w_2 + 2w_3 - 2w_4 + 7w_5 \leq -2$$

$$4w_1 + 3w_2 + w_3 + 5w_4 - 2w_5 \leq 4$$

$$w_1, w_2, w_3, w_4, w_5 \geq 0$$

Multiply the second constraint by

$$-1 - 5w_1 - w_2 - 2w_3 + 2w_4 - 7w_5 \geq 2$$

Introducing the slack variable w_{1s} and w_{3s} in the first and third constraint introducing and surplus variable and artificial variable w_{2s} an w_{29} in the second constraint, and taking the negative price M to artificial variable, the dual problem reduce to the following form.

$$\text{Max} \quad Z_D = 7w_1 + 4w_2 - 10w_3 + 3w_4 + 2w_5$$

$$\text{s.t.} \quad 3w_1 + 6w_2 - 7w_3 + w_4 + 4w_5 + w_{1s} = 3$$

$$-5w_1 - w_2 - 2w_3 + 2w_4 - 7w_5 - w_{2s} + w_{2a} = 2$$

$$4w_1 + 3w_2 + w_3 + 5w_4 - 2w_5 + w_{3s} = 4$$

$$w_1, w_2, w_3, w_4, w_5, w_{1s}, w_{2s}, w_{3s}, w_4 \geq 0$$

Taking $w_1 = 0 = w_2 = w_3 = w_4 = w_5 = w_{2s}$, we have $w_{1s} = 3$, $w_{2a} = 2$, $w_{3s} = 4$. Which is the starting basic feasible solution of the dual.

First Simplex Table

		c_j	7	4	−10	3	2	0	0	0	−M	w_B
B	c_B	w_B	Y_1	Y_2	Y_3	Y_4	Y_5	Y_{1s}	Y_{2s}	Y_{3s}	Y_{2a}	y_4
x_{1s}	0	3	3	6	−7	1	4	1	0	0	0	3/1
x_{29}	−M	2	−5	−1	−2	2	−7	0	−1	0	1	2/2
x_{3s}	0	4	4	3	1	[5]	−2	0	0	1	0	4/8mini
		$Z_j - c_j s_M$	−7M−4	10+2_M	−3-2M	7M−2	0	M	0	0		
					↑				↓			

Since $\Delta_4 = -3 - 2_M$ is mini of all. So y_4 is incoming vector and by mini ratio rule out going vector is y_{3s}.

The key element is $y_{34} = 5$

∴ Proceeding as usual the second simplex table.

Second Table

B	c_B	c_j w_B	7 Y_1	4 Y_2	–10 Y_3	3 Y_4	2 Y_5	0 Y_{1s}	0 Y_{2s}	0 Y_{3s}	–M Y_{2a}	$\frac{w_B}{y}$
x_{1s}	0	11/5	11/5	27/5	–36/5	22/5	0	1	0	–1/5	0	
x_{2a}	–M	2/5	–33/5	–11/5	–12/5	0	–31/5	0	–1	–2/5	1	
x_4	3	4/5	4/5	3/5	1/5	1	–2/5	0	0	1/5	0	
	$Z_j - c_j$	$\frac{33M-2s}{5}$	$\frac{11M-11}{5}$		$\frac{53+12M}{5}$	0	$\frac{31M-16}{5}$	0	M	$\frac{2M+3}{5}$	0	

Since all $Z_j - c_j \geq 0$, so this solution is optimal but the artificial vector, appear in the basis at the positive level which indicates that the dual problem has no feasible solution, so primal also has no solution.

Example 6:

Apply the principle of duality to solve the following linear programming problem.

$$\text{Min. } Z_D = 2x_1 + 2x_2$$

$$\text{s.t. } 2x_1 + 4x_2 \geq 1$$

$$2x_1 + x_2 \geq 1$$

$$x_1 + 2x_2 \geq 1$$

$$x_1, x_2 \geq 0.$$

Solution:

The dual of this problem is given by

$$\text{Max. } Z_D = w_1 + w_2 + w_3$$

$$\text{s.t. } 2w_1 + 2w_2 + w_3 \leq 2$$

$$4w_1 + w_2 + 2w_3 \leq 3$$

Introducing the slack variable w_{1s} and w_{2s}. The dual problem can be written as

$$\text{Max. } Z_D = w_1 + w_2 + w_3$$

$$\text{s.t. } 2w_1 + 2w_2 + w_2 + w_{1s} = 2$$

$$4w_1 + w_2 + 2w_3 + w_{2s} = 3$$

$$w_1, w_2, w_{1s}, w_{2s} \geq 0$$

Taking $w_1 = 0 = w_2 = w_3$, we have $w_{1s} = 2$ $w_{2s} = 2$. Which is the starting basic feasible solution of the dual.

First Simplex Table

B	c_B	c_J / w_B	1 Y_1	1 Y_2	1 Y_3	0 Y_{1s}	0 Y_{2s}	$\frac{w_B}{Y_1}$
x_{1s}	0	2	2	2	1	1	0	1
x_{2s}	0	2	[4]	1	2	0	1	1/2 mini
		$Z_J - c_J$	–1	–1	–1	0	0	

Proceeding as usual the second simplex table.

Second Table

B	c_B	c_J / w_B	1 Y_1	1 Y_2	1 Y_3	0 Y_{1s}	0 Y_{2s}	$\frac{w_B}{y}$
x_{1s}	0	1	0	3/2	0	1	–1/2	x
x_1	1	1/2	1	1/4	[1/2]	0	1/4	1 mini
			0	–3/4	–1/2	0	1/4	

Proceeding as usual third simplex table.

Third Table

B	c_B	c_J / w_B	1 Y_1	1 y_2	1 Y_3	0 Y_{1s}	0 Y_{2s}	$\frac{w_B}{y_2}$
x_{1s}	0	1	0	[3/2]	0	1	–1/2	2/3 Mini
x_3	1	1	2	1/2	1	0	1/2	2
		$Z_J - c_J$	1	–1/2	0	0	1/2	
				↑		↓		

Proceeding as usual fourth simplex table.

Fourth Table

B	c_B	c_J / w_B	1 Y_1	1 Y_2	1 Y_3	0 Y_{1s}	0 Y_{2s}
x_2	1	2/3	0	1	0	2/3	–1/3
x_3	1	2/3	2	0	1	–1/3	2/3
		$Z_J - c_J$	1	0	0	1/3	1/3

Since all $Z_J - c_J \geq 0$. So solution is optimal.

Solution of primal problem

$x_1 = 1/3$

$x_2 = 1/3$

$Z = 4/3$

For primal constraints, read across the table while for dual constraints read down the columns.

Example 7:

Write the dual and solve the primal problem by simplex method.

$$\text{Max. } Z = 2x_1 - x_2$$

$$\text{s.t.} \quad x_1 + x_2 \le 10$$

$$4x_1 + 3x_2 \ge 12$$

$$-2x_1 + x_2 \le 2$$

$$x_1, x_2 \ge 0.$$

Solution:

In standard form the problem can be written as

$$\text{Max. } Z = 2x_1 - x_2$$

$$\text{s.t.} \quad x_1 + x_2 \le 10$$

$$-(4x_1 + 3x_2) \le -12$$

$$-2x_1 + x_2 \le 2$$

$$x_1, x_2 \ge 0$$

The dual of this problem can be written as

$$\text{Mini. } Z_D = 10w_1 - 12w_2 + 2w_2$$

$$\text{s.t.} \quad w_1 - 4w_2 - 2w_3 \ge 2$$

$$w_1 - 3w_2 + w_2 \ge -1$$

Introducing the slack variable in first and third constraint and introducing surplus and artificial variable is second constraint of the primal problem.

$$\text{Max. } Z = 2x_1 - x_2$$

$$\text{s.t.} \quad x_1 + x_2 + x_{1s} = 10$$

$$4x_1 + 3x_2 - x_{2s} + x_{2a} = 12$$

$$-2x_1 + x_2 + x_{3s} = 2$$

Taking $x_1 = 0 = x_2 = x_{2s}$, we have $x_{1s} = 10$, $x_{2a} = 12$ $x_{3s} = 2$ which is the starting basic feasible solution of the primal

First Table

B	c_B	c_J / x_B	2 / Y_1	–1 / Y\	0 / Y_{1s}	0 / Y_{2s}	0 / Y_{3s}	–M / Y_{2a}	x_B / Y_1
Y_{4s}	0	10	1	1	1	0	0	0	10
x_{2a}	–M	12	[4]	3	0	–1	0	1	3mini
x_{3s}	0	2	–2	1	0	0	1	0	–ve
		Z_J–c_J	–4M–2 ↑	–3m+1	0	M	0	0 ↓	

Since $\Delta_1 = -4_m - 2$ is mini so y_1 is incoming vector and mini ratio rule out going vector is y_{2a}.

The key element is $y_{21} = 4$

Proceeding as usual the second simplex table.

Second Table

B	c_B	c_J / x_B	2 / Y_1	–1 / Y_2	0 / Y_{1s}	0 / Y_{2s}	0 / Y_{3s}	–M / Y_{2a}	y_B / Y_{2s}
x_{1s}	0	7	0	1/4	1	[1/4]	0	–1/4	21
x_1	2	3	1	3/4	0	–1/4	0	1/4	–ve
x_{3s}	0	8	0	5/2	0	–1/2	1	1/2	–ve
		Z_J–c_J	0	5/2	0	–1/2 ↑	0	$\frac{1+2M}{2}$	

Since $\Delta_4 = -1/2$ is mini so y_{2s} is incoming vector and mini ratio rule out going vector is y_{1s}.

The key element is $y_{14} = 1/4$

Proceeding as usual the third simplex table

Third Table

B	c_J / c_B	2 / x_B	–1 / Y_1	0 / Y_2	0 / Y_{1s}	0 / Y_{2s}	–M / Y_{3s}	Y_{2a}
x_{2s}	0	28	0	1	4	1	0	–1
x_1	2	10	1	1	1	0	0	0
x_{3s}	0	22	0	3	2	0	1	0
		Z_J–c_J	0	1	2	0	0	0

Since all $Z_j - c_j \geq 0$, so solution is optimal

$$x_1 = 10, x_2 = 0$$

Max. Z = 20.

Example 8:

Find the dual of problem

$$Max.\ Z = -x_1 + 2x_2 - x_3$$

$$s.t.\quad 3x_1 + x_2 - x_3 \leq 10$$

$$-x_1 + 4x_2 + x_3 \geq 6$$

$$x_2 + x_3 \leq 4$$

$$x_1, x_2, x_3 \geq 0$$

Solve the primal problem by simplex method and deduce the solution of the problem from the optimal tableaus of the primal.

Solution:

In standard form the problem can be written as

$$\text{Max. } Z_P = -x_1 + 2x_2 - x_3$$

$$3x_1 + x_2 \quad x_3 \leq 10$$

$$x_2 - 4x_2 - x_3 \leq 6$$

$$x_2 + x_3 \leq 4$$

$$x_1, x_2, x_3, x_4 \geq 0$$

The dual of this problem can be written as

$$\text{Mini. } Z_D = 10w_1 + 6w_2 + 4w_3$$

$$3w_1 \geq -1$$

$$w_1 + w_2 + w_3 \geq 2$$

$$-w_1 - w_2 + w_3 \geq -1$$

Introducing the slack variable in first and third constraint and surplus and artificial is second constraint of the primal problem. The problem can be written as

$$3x_1 + x_2 - x_3 + x_{1s} = 10$$

$$-x_1 + 4x_2 + x_3 - x_{2s} + x_{2a} = 6$$

$$x_2 + x_3 + x_{3s} = 4$$

$$x_1, x_2, x_3, \ldots x_{3s} \geq 0$$

Taking $x_1 = 0 = x_2 = x_3 = x_{2s}$, we have $x_{1s} = 10$, $x_{2a} = 6$, $x_{3s} = 4$, which is the starting basic feasible solution of the primal.

First Table

B	c_B	c_J / x_B	−1 Y_1	2 Y_2	−1 Y_3	0 Y_{1s}	0 Y_{2s}	0 Y_{3s}	−M Y_{2a}	x_B/Y_2
x_{1s}	0	10	3	1	−1	1	0	0	0	10
x_{2a}	−M	6	−1	[4]	1	0	−1	0	1	3/2 mini→
x_{3s}	0	4	0	1	1	0	0	1	0	4
		Z_J-c_J	M+1	−4+2 ↑	−M+1	0	M	0	0 ↓	

Since $\Delta_2 = -4M - 2$ is mini so y_2 is incoming vector and by mini ratio rule out going vector is y_{2a}.

The key element is $y_{22} = 4$

Proceeding as before the second simplex table

Second Table

B	c_B	c_J / x_B	−1 Y_1	2 Y_2	1 Y_3	0 Y_{1s}	0 Y_{2s}	0 Y_{3s}	−M Y_{2a}	x_B/Y_{2s}
x_{1s}	0	17/2	13/4	0	−5/4	1	1/4	−1/4	0	34
x_2	2	3/2	−1/4	1	1/4	0	−1/4	1/4	0	−ve
x_{3s}	0	5/2	1/4	0	3/4	0	1/4	−1/4	1	10 mini
		Z_J-c_J	1/2	0	11/8	0	−1/2	$\frac{1}{2}$ + M		0

Proceeding as before the third simplex table

Third Table

B	c_B	c_J / x_B	−1 Y_1	2 Y_2	−1 Y_3	0 Y_{1s}	0 Y_{2s}	0 Y_{3s}	−M Y_{2a}
x_{1s}	0	29/4	3	0	−2	1	0	−1	0
x_2	2	4	0	1	1	0	0	1	0
x_{2s}	0	10	1	0	3	0	1	4	−1
		$Z_J - c_J$	1	0	3	0	0	2	M

Since all $Z_J - c_J \geq 0$, so this solution is optimal.

Optimal solution is

$$x_1 = 0 = x_3, \; x_2 = 4$$

$$\text{Max. } Z = 8$$

solution of the dual

$$w_1 = 0 = w_2, \; w_3 = 2$$

$$\text{Mini. } Z = 8.$$

Example 9:

Consider the problem,

$$\text{Max. } Z = 8x_1 + 6x_2$$

$$\text{s.t.} \quad x_1 - x_2 \leq 3/5$$

$$x_1 - x_2 \geq 2$$

$$x_1, x_2 \geq 0$$

Show that the primal and dual problem have no feasible solution.

Solution:

In standard form the problem can be written as

$$\text{Max. } Z = 8x_1 + 6x_2$$

$$\text{s.t.} \quad x_1 - x_2 \leq 3/5$$

$$-x_1 + x_2 \leq -2$$

$$x_1, x_2 \geq 0$$

The dual of the this problem can be written as

$$\text{Mini. } Z_D = 3/5\, w_1 - w_2$$

$$w_1 - w_2 \geq 8$$

$$-w_1 + w_2 \geq 6$$

$$w_1, w_2 \geq 0$$

Introducing the slack variable in fist constraint and surplus and artificial variable in second constraint of the primal.

The problem can be written as

$$\text{Max. } Z = 8x_1 + 6x_2$$

$$\text{s.t.} \quad x_1 - x_2 + x_{1s} = 3/5$$

$$x_1 - x_2 - x_{1s} + x_{2a} = 2$$

$$x_1, \ldots, x_{2a} \geq 0$$

Taking $x_1 = x_2 = x_{2s}$ we have $x_{1s} = 3/5$ $x_{2a} = 2$. Which is starting basic feasible solution of the primal.

First Table

B	c_B	c_J / x_B	8 / Y_1	6 / Y_2	0 / Y_{1s}	0 / Y_{2s}	–M / Y_{2a}
x_{1s}	0	3/5	[1]	–1	1	0	0
x_{2a}	–M	2	1	–1	0	–1	1
		$Z_J–c_J$	–M–8	+M–6	0	M	0

Proceeding as before the second simplex table.

Second Table

B	c_B	c_J / x_B	8 / Y_1	6 / Y_2	0 / Y_{1s}	0 / Y_{2s}	–M / Y_{2a}	x_B / Y_2
c_1	8	3/5	1	–1	1	0	0	–ve
x_{2a}	–M	7/5	0	0	–1	–1	1	∞
	$Z_J–c_J$	0	–2 ↑	M+8	M	0		

Every value of $\frac{x_B}{y_2}$ are –ve and ∞, so the primal has no feasible solution and dual has also no feasible solution.

Example 10:

Write the dual of this problem

$Max.\ Z = 2x_1 + 3x_2$

$s.t.\quad 2x_1 + 2x_2 \le 10$

$x_1 + 2x_2 \le 6$

$2x_1 + x_2 \le 6$

$x_1, x_2 \ge 0.$

Solve the primal by the simplex method.

Solution:

The due of this problem can be written as

$\text{Mini. } Z_D = 10w_1 + 6w_2 + 6w_3$

$2w_1 + w_2 + 2w_3 \ge 2$

$2w_1 + 2w_2 + w_3 \geq 3$

$w_1, \ldots, w_3 \geq 0$

Introducing the slack variable x_{1s}, x_{2s}, x_{3s} in the primal problem. The problem can be written as

Max. $Z = 2x_1 + 3x_2$

s.t. $2x_1 + 2x_2 + x_{1s} = 10$

$x_1 + 2x_2 + x_{2s} = 6$

$2x_1 + x_2 + x_{3s} = 6$

$x_1, \ldots, x_{3s} \geq 0$

Taking $x_1 = 0 = x_2$ have $x_{1s} = 10, x_{2s} = 6, x_{3s} = 6$

Which is the starting basic feasible solution of the primal.

First Table

		c_J	2	2	0	0	0	x_B
B	c_B	x_B	Y_1	Y_2	Y_{1s}	Y_{2s}	Y_{3s}	Y_1
x_{1s}	0	10	2	2	1	0	0	5
x_{2s}	0	6	1	2	0	1	0	6
x_{3s}	0	6	[2]	1	0	0	1	2 mini
	$Z_J - c_J$		–2	–2	0	0	0	
			↑				↓	

Proceeding as before the second simplex table.

Second Table

		c_J	2	2	0	0	0	x_B
B	c_B	x_B	Y_1	Y_2	Y_{1s}	Y_{2s}	Y_{3s}	Y_2
x_{1s}	0	4	0	1	1	0	–1	4
x_{2s}	0	3	0	[3/2]	0	1	–1/2	6
x_1	2	3	1	1/2	0	0	1/2	6
			0	–1	0	0	1	
				↑		↓		

Proceeding as before the third simplex table.

Third Table

B	c_B	c_J x_B	Y_1	Y_2	Y_{1s}	Y_{2s}	Y_{3s}
x_{1s}	0	2	0	2	1	2/3	–2/3
x_2	3	2	0	1	0	2/3	–1/3
x_1	2	2	1	0	0	–1/3	2/3
	$Z_J–c_J$		0	1	0	4/3	1/3

Since all $Z_J - c_J \geq 0$, so this solution is optimal. The solution of the primal problem is

$x_1 = 2 = x_2$

Max $Z_P = 10$

The solution of dual $w_1 = 0$

$w_2 = 4/3$

$w_3 = 1/3$

Mini. $Z_D = 10$

Example 11:

Use duality to solve the L.P.P.

Max. $Z = 5x_1 - 2x_2 + 3x_3$

s.t. $2x_1 + 2x_2 - x_3 \geq 2$

$3x_1 - 4x_2 \leq 3$

$x_2 + 2x_3 \geq 5$

$x_1, x_2, x_3 \geq 0.$

Solution:

In the standard form the problem is written as

Max. $Z = 5x_1 - 2x_2 + 3x_3$

s.t. $2x_1 - 2x_2 + x_3 \leq -2$

$3x_1 - 4x_2 \leq 3$

$x_2 + 3x_3 \leq 5$

$x_1, x_2, x_3 \leq 0$

The dual of this problem is given by

Min $Z_D = -2w_1 + 3w_2 + 5w_3$

$$\text{s.t.} \quad -2w_1 + 3w_2 \geq 5$$

$$-2w_1 - 4w_2 + w_3 \geq -2$$

$$w_1 + 3w_3 \geq 3$$

$$w_1, w_2, w_3 \geq 0$$

Multiply the second constraint by – 1

$$2w_2 + 4w_2 - w_3 \leq 2$$

Introducing the surplus and artificial variable in first and third constraint and slack variable in second constraint. The dual problem reduce to the following form.

$$\text{Max. } Z_D = 2w_1 - 3w_2 - 5w_3$$

$$\text{s.t.} \quad -2w_1 + 3w_2 - w_{1s} + w_{1a} = 5$$

$$2w_1 + w_2 - w_3 + w_{2s} = 2$$

$$w_1 + 3w_3 - w_{3s} + w_{3a} = 3$$

$$w_1, w_2, \ldots, w_{3a} \geq 0$$

Taking $w_1 = 0 = w_2 = w_3 = w_{1s} = w_{3s}$, we have $w_{1a} = 5$, $w_{2s} = 2$, $w_{3a} = 3$

Which is the starting basic feasible solution of the dual.

First Table

B	c_B	c_j / w_B	2 / Y_1	–3 / Y_2	–5 / Y_3	0 / Y_{1s}	0 / Y_{2s}	0 / Y_{3s}	–M / Y_{1a}	–M / Y_{3a}	$\frac{w_B}{Y_2}$
x_{1a}	–M	5	–2	3	0	–1	0	0	1	0	5/3
x_{2s}	0	2	2	[4]	–1	0	1	0	0	0	2/4mini→
x_{3a}	–M	3	1	0	3	0	0	–1	0	1	∞
		$Z_j - c_j$	M–2	–3M+3 ↑	–3M+5	M	0 ↓	M	0	0	

Proceeding as before second simplex table

Second Table

B	c_B	c_j / w_B	2 / Y_1	–3 / Y_2	–5 / Y_3	0 / Y_{1s}	0 / Y_{2s}	0 / Y_{3s}	–M / Y_{1a}	–M / Y_{2a}	$\frac{w_B}{Y_3}$
x_{1a}	–M	7/2	–7/2	0	3/4	–1	–3/4	0	1	0	14/3
x_2	–3	1/2	1/	1	–1/4	0	1/4	0	0	0	–ve
x_{3s}	–M	3	1	0	[3]	0	0	–1	0	1	1 mini
		$Z_j - c_j$	$\frac{5M-7}{2}$	0	$\frac{23-15M}{2_1}$ ↑	M	$\frac{3M-3}{4}$	M	0	0 ↓	

Proceeding as usual third simplex table

Third Table

B	c_B	c_J / w_B	2 / Y_1	–3 / Y_2	–5 / Y_3	0 / Y_{1s}	0 / Y_{2s}	0 / Y_{3s}	–M / Y_{1a}	–M / Y_{3a}	w_B
x_{1a}	–M	11/4	–15/4	0	0	–1	–3/4	(1/4)	1	–1/4	11 mini.
x_2	–3	3/4	7/12	1	0	0	1/4	–1/12	0	1/12	–ve
x_3	–5	1	1/3	0	1	0	0	–1/3	0	1/3	–ve
			$\frac{45M-95}{12}$	0	0	–M	$\frac{3M-3}{4}$	0	$\frac{5M-6}{4}$		
							↑		↓		

Proceeding as usual fourth simplex table

Fourth Table

B	c_B	c_J / w_B	2 / y_1	–3 / Y_2	–5 / Y_3	0 / Y_{1s}	0 / Y_{2s}	0 / Y_{3s}
x_{3s}	0	11	–15	0	0	–4	–3	1
x_2	–3	5/3	–2/3	1	0	–1/3	0	0
x_3	–5	14/3	–14/3	0	1	–4/3	–1	0
		$Z_J - c_J$ 0	0	0	23/3	5	0	

Since all $Z_J - c_J \geq 0$, so the solution is optimal.

∴ Solution of the primal problem is

$x_1 = 23/3 \;\; x_3 = 0$

$x_2 = 5$

$\text{Max. } Z_P = \frac{85}{3}.$

EXERCISE

1. Define the dual of given L.P.P. prove that dual of a dual is primal problem.
2. Prove that if the primal has an unbounded solution, the dual has no solution and vice-versa.
3. Use the duality to solve the following problem.

   (i) Max $Z = 3x_1 + x_2$

   s.t. $x_1 + x_2 \leq 1$

$2x_1 + 3x_2 \geq 2$

$x_1, x_2 \geq 0.$

(ii) Mini. $Z_P = x_1 + x_2 + x_3$

$x_1 + 3x_2 + 4x_3 = 5$

$x_1 - 2x_2 \leq 3$

$2x_2 - x_3 \geq 4$

$x_1, x_2 \geq 0$

x_3 is unrestricted.

(iii) Max $Z = 4x_1 + 3x_2$

$x_1 \leq 6$

$x_2 \leq 8$

$x_1 + x_2 \leq 7$

$3x_1 + x_2 \leq 15$

$-x_2 \leq 1$

$x_1, x_2 \geq 0.$

(iv) Mini $Z = 3x_1 - 2x_2 + 4x_3$

s.t. $3x_1 + 5x_2 + 4x_3 \geq 7$

$6x_1 + x_2 + 3x_3 \geq 4$

$7x_1 - 2x_2 - x_3 \leq 10$

$x_1 - 2x_2 + 5x_3 \geq 3$

$4x_1 + 7x_2 - 2x_3 \geq 2$

$x_1, x_2, x_3 \geq 0.$

(v) Mini. $Z = 4x_1 + 3x_2 + 6x_3$

s.t. $x_1 + x_3 \geq 2$

$x_2 + x_3 \geq 5$

$x_1, x_2, x_3 \geq 0.$

(vi) Max. $Z = 2x_1 + 3x_2 + 5x_3$

s.t. $x_1 + x_2 + x_3 \leq 7$

$3x_1 - x_2 + x_3 \leq 5$

$x_1 + 2x_2 + 2x_3 \leq 13$

$x_1, x_2, x_3 \geq 0.$

ANSWERS

3.

(i) $x_1 = 0$

$x_2 = 1$

Mini. $Z = 1$

(iii) $x_1 = 4$

$x_2 = 3$

Max. $Z = 25$

(iv) No. F.S

(v) $x_1 = 0, x_2 = 3, x_3 = 2$

Mini. $Z = 21$.

UNSYMMETRIC PRIMAL DUAL PROBLEMS

Primal Problem

Find a column vector $\mathbf{x} \in R^n$ which maximizes

$$Z_x = c\,x,\ c \in R^n$$

subject to $A\,x = b,\ b \in R^m$,

$$x \geq 0 \text{ and } A \text{ is an } m \times n \text{ real matrix.}$$

Dual Problem

Find a column vector $w \in R^m$ which minimizes

$$Z_w = b'\,w$$

subject to $A'\,w \geq c'$.

In this case *the dual variables are unrestricted in sign.*

DUAL OF AN L.P.P. WITH MIXED RESTRICTIONS

Sometimes a given L.P.P. contains a mixture of inequalities ($\geq$, $\leq$), equations; non-negative variables and unrestricted variables, then to obtain its dual we proceed in the following manner:

(i) If a constraint is an equation (has = sign), replace it by two constraints involving the inequalities going in opposite directions.

For example,

the equation $2x_1 + 5x_2 = 9$ is replaced by

$$2x_1 + 5x_2 \leq 9 \qquad ...(1)$$

and $2x_1 + 5x_2 \geq 9$. ...(2)

(ii) If the given problem is of maximization, all constraints should have $\leq$ sign. If some constraint has $\geq$ sign, multiply both sides by -1 and make the sign $\leq$.

Thus in the above example multiply both sides of (2) by -1 so that it becomes

$-2x_1 - 5x_2 \leq -9$

Similarly, if the problem is of minimization, all constraints should have $\geq$ sign.

(iii) If there is some unrestricted variable, replace it by the difference of two variables.

Note: The dual variables corresponding to primal equality constraints must by unrestricted in sign and those associated with primal inequalities must be non-negative.

Standard Primal Form

A linear programming problem is said to be in *standard primal form if*

(a) For a maximization problem all the constraints have $\leq$ sign.

(b) For a minimization problem all the constraints have $\geq$ sign.

Example 1:

Write the dual of the following problem

$$\text{Min. } Z = 10x_1 + 20x_2$$

subject to $3x_1 + 2x_2 \geq 18$

$$x_1 + 3x_2 \geq 8$$

$$2x_1 - x_2 \leq 6$$

$$x_1, x_2 \geq 0.$$

Solution:

First we shall convert the given L.P.P. into standard primal form.

1. Since the given problem is of minimization, therefore all the constraints should have $\geq$ sign.
2. The first two inequalities are in the right direction while the third one is not.
3. Multiplying both sides of the third inequality by -1 it changes to $-2x_1 + x_2 \geq -6$.

4. Thus the standard primal form of the given L.P.P. is

Min $Z = 10x_1 + 20x_2$

subject $3x_1 + 2x_2 \geq 18$

$x_1 + x_2 \geq 8$

$-2x_1 + x_2 \geq 6.$

and $x_1, x_2 \geq 0.$

5. The matrix form of the above L.P.P. is

Min. $Z = (10, 20)\ (x_1, x_2)$

$= c\,x$

subject

$$\begin{bmatrix} 3 & 2 \\ 1 & 3 \\ -2 & 1 \end{bmatrix} \begin{bmatrix} x_1 \\ x_2 \end{bmatrix} \geq \begin{bmatrix} 18 \\ 8 \\ -6 \end{bmatrix}$$

or $A\,x \geq b,$

$x_1, x_2 \geq 0.$

$\therefore$ the required dual is

Max. $Z_D = b'\,y = (18, 8, -6)\ (y_1, y_2, y_3)$

s.t. $A'\,y = c'$ or $\begin{bmatrix} 3 & 1 & -2 \\ 2 & 3 & 1 \end{bmatrix} \begin{bmatrix} y_1 \\ y_2 \\ y_3 \end{bmatrix} \leq \begin{bmatrix} 10 \\ 20 \end{bmatrix}$

or Max. $Z_D = 18y_1 + 8y_2 - 6y_3$

s.t. $3y_1 + y_2 - 2y_3 \leq 10$

$2y_1 + 3y_2 + y_3 \leq 20,$

$y_1, y_2, y_3 \geq 0.$

Example 2:

Write the dual of the following problem:

Min. $Z = x_1 + x_2 + x_3$

subject to $x_1 - 3x_2 + 4x_3 = 5,$

$x_1 - 2x_2 \leq 3,\ 2x_2 - x_3 \geq 4,$

$x_1, x_2 \geq 0,\ x_3$ *is unrestricted.*

Solution:

First we shall write the given L.P.P. in the standard primal form, substituting $x_3 = x_3' - x''$, $x_3' \geq 0$, $x_3' \geq 0$. The given problem can be written in the standard primal form as

$$\text{Min. } Z = x_1 + x_2 + x_3' - x_3$$

subject to $- x_1 + 3x_2 - 4\,(x_3' - x_3'') \geq -5$

$$x_1 - 3x_2 + 4\,(x_3' - x_3'') \geq 5$$

$$-x_1 + 2x_2 \geq -3$$

$$2x_2 - (x_3' - x_3'') \geq 4,$$

$$x_1, x_2, x_3', x_3'' \geq 0.$$

The matrix form of the above problem is

$$\text{Min. } Z = (1, 1, 1, -1)\,(x_1, x_2, x_3', x_3'')$$

subject to

$$\begin{bmatrix} -1 & 3 & -4 & 4 \\ 1 & -3 & 4 & -4 \\ -1 & 2 & 0 & 0 \\ 0 & 2 & -1 & 1 \end{bmatrix} \begin{bmatrix} x_1 \\ x_2 \\ x_3' \\ x_3'' \end{bmatrix} \geq \begin{bmatrix} -5 \\ 5 \\ -3 \\ 4 \end{bmatrix}$$

Now the dual of the given primal is

$$\text{Max } Z_D = (-5, 5, -3, 4)\,(y_1', y_1'', y_2, y_3)$$

$$= -5\,(y_1' - y_1'') - 3y_2 + 4y_3$$

subject to

$$\begin{bmatrix} -1 & 1 & -1 & 0 \\ 3 & -3 & 2 & 2 \\ -4 & 4 & 0 & -1 \\ 4 & -4 & 0 & 1 \end{bmatrix} \begin{bmatrix} y_1' \\ y_1'' \\ y_2 \\ y_3 \end{bmatrix} \leq \begin{bmatrix} 1 \\ 1 \\ 1 \\ -1 \end{bmatrix}$$

or

$$-(y_1' - y_1'') - y_2 \leq 1$$

$$3\,(y_1' - y_1'') + 2y_2 + 2y_3 \leq 1$$

$$-4\,(y_1' - y_1'') - y_3 \leq 1$$

$$4\,(y_1' - y_1'') + y_3 \leq -1,$$

$$y_1', y_1'', y_2, y_3 \geq 0.$$

Substituting $y_1 = y_1' - y_1''$, the required dual is

$$\text{Max. } \quad Z_D = -5y_1 - 3y_2 + 4y_3$$

subject to $-y_1 - y_2 \leq 1$

$3y_1 + 2y_2 + 2y_3 \leq 1$

$-4y_1 - y_3 \leq 1$

$4y_1 + y_3 \leq -1$ or $-4y_1 - y_3 \geq 1$,

$y_2, y_3 \geq 0$ and y_1 is unrestricted.

Example 3:

Write the dual of the following problem:

Min. $Z = 2x_2 + 5x_3$

subject $x_1 + x_2 \geq 2$

$2x_1 + x_2 + 6x_3 \leq 6$

$x_1 - x_2 + 3x_3 = 4$

$x_1, x_2, x_3 \geq 0.$

Solution:

First we shall convert the given problem to standard primal form:

1. Since the problem is of minimization therefore all the constraints should have the sign $\geq$.
2. Multiplying the second constraint by -1, it becomes $-2x_1 - x_2 - 6x_3 \geq -6$.
3. Since the third constraint is an equality so replacing it by the following two constrains

$$x_1 - x_2 + 3x_3 \leq 4$$

$$x_1 - x_2 + 3x_3 \geq 4$$

or $$-x_1 + x_2 - 3x_3 \geq -4$$

and $$x_1 - x_2 + 3x_3 \geq 4.$$

$\therefore$ the given problem in standard primal form is

Min. $Z = 0x_1 + 2x_2 + 5x_3$

subject to $x_1 + x_2 \geq 2$

$-2x_1 - x_2 - 6x_3 \geq -6$

$-x_1 + x_2 - 3x_3 \geq -4$

$x_1 - x_2 + 3x_3 \geq 4$

$x_1, x_2, x_3 \geq 0.$

The matrix form of the above problem is

Min. $Z = (0, 2, 5)\ (x_1, x_2, x_3)$

subject to

$$\begin{bmatrix} 1 & 1 & 0 \\ -2 & -1 & -6 \\ -1 & 1 & -3 \\ 1 & -1 & 3 \end{bmatrix} \begin{bmatrix} x_1 \\ x_2 \\ x_3 \end{bmatrix} \geq \begin{bmatrix} 2 \\ -6 \\ -4 \\ 4 \end{bmatrix}.$$

Now the dual of the given primal is

Max. $Z_D = (2, -6, -4, 4)\ (y_1, y_2, y_3', y_3'')$

subject to

$$\begin{bmatrix} 1 & -2 & -1 & 1 \\ 1 & -1 & 1 & -1 \\ 0 & -6 & -3 & 3 \end{bmatrix} \begin{bmatrix} y_1 \\ y_2 \\ y_3' \\ y_3'' \end{bmatrix} \leq \begin{bmatrix} 0 \\ 2 \\ 5 \end{bmatrix},$$

$$y_1, y_2, y_3', y_3'' \geq 0$$

or Max. $Z_D = 2y_1 - 6y_2 - 4\ (y_3' - y_3'')$

subject to

$$y_1 - 2y_2 - (y_3' - y_3'') \leq 0$$

$$y_1 - y_2 + y_3' - y_3'' \leq 2$$

$$-6y_2 - 3\ (y_3' - y_3'') \leq 5,$$

$$y_1, y_2, y_3', y_3'' \geq 0.$$

Substituting $y_3 = y_3' - y_3''$, the required dual is

Max. $Z_D = 2y_1 - 6y_2 - 4y_3$

subject to

$$y_1 - 2y_2 - y_3 \leq 0$$

$$y_1 - y_2 + y_3 \leq 2$$

$$-6y_2 - 3y_3 \leq 5,$$

$y_1, y_2 \geq 0$ and y_3 is unrestricted in sign.

Example 4:

Write the dual of the following problem

Min. $Z = 3x_1 - 2x_2 + 4x_3$

subject to $3x_1 + 5x_2 + 4x_3 \geq 7$

$6x_1 + x_2 + 3x_3 \geq 4$

$7x_1 - 2x_2 - x_3 \leq 10$

$x_1 - 2x_2 + 5x_3 \geq 3$

$4x_1 + 7x_2 - 2x_3 \geq 2$

$x_1, x_2, x_3 \geq 0.$

Solution:

First we shall convert the given L.P.P. into standard prima form.

Since the given problem is of minimization therefore all the constraints should have the sign ≥ .

Thus the standard primal form of the given L.P.P. is

Min $Z = 3x_1 - 2x_2 + 4x_3$

subject to $3x_1 + 5x_2 + 4x_3 \geq 7$

$6x_1 + x_2 + 3x_3 \geq 4$

$-7x_1 + 2x_2 + x_3 \geq -10$

$x_1 - 2x_2 + 5x_3 \geq 3$

$4x_1 + 7x_2 - 2x_3 \geq 2, x_1,$

$x_2, x_3 \geq 0.$

The matrix form of the above problems

Min $Z = (3, -2, 4)\ (x_1, x_2, x_3)$

$$\text{subject to} \begin{bmatrix} 3 & 5 & 4 \\ 6 & 1 & 3 \\ -7 & 2 & 1 \\ 1 & -2 & 5 \\ 4 & 7 & -2 \end{bmatrix} \begin{bmatrix} x_1 \\ x_2 \\ x_3 \end{bmatrix} \geq \begin{bmatrix} 7 \\ 4 \\ -10 \\ 3 \\ 2 \end{bmatrix}.$$

Now the dual of the given primal is

Max. $Z_D = (7, 4, -10, 3, 2)\ (y_1, y_2, y_3, y_4, y_5)$

subject to

$$\begin{bmatrix} 3 & 6 & -7 & 1 & 4 \\ 5 & 1 & 2 & -2 & 7 \\ 4 & 3 & 1 & 5 & -2 \end{bmatrix} \begin{bmatrix} y_1 \\ y_2 \\ y_3 \\ y_4 \\ y_5 \end{bmatrix} \leq \begin{bmatrix} 3 \\ -2 \\ 4 \end{bmatrix}$$

or Max. $Z_D = 7y_1 + 4y_2 - 10y_3 + 3y_4 + 2y_5$

subject to $3y_1 + 6y_2 - 7y_3 + y_4 + 4y_5 \leq 3$

$$5y_1 + y_2 + 2y_3 - 2y_4 + 7y_5 \le 3$$
$$4y_1 + 3y_2 + y_3 + 5y_4 - 2y_5 \le 4,$$
$$y_1, y_2, y_3, y_4, y_5 \ge 0.$$

Example 5:

Write the dual of the following L.P.P.

$$\text{Max. } Z = 2x_1 + 3x_2 + x_3$$
subject to $4x_1 + 3x_2 + x_3 = 6$
$$x_1 + 2x_2 + 5x_3 = 4,$$
$$x_1, x_2, x_3 \ge 0.$$

Solution:

First we shall convert the given L.P.P. into standard primal form.

Since the given problem is of maximization, so all the constraints should have the sign $\le$.

The standard primal form of the given L.P.P. is

$$\text{Max. } Z = 2x_1 + 3x_2 + x_3$$
subject to $4x_1 + 3x_2 + x_3 \le 6$
$$-4x_1 - 3x_2 - x_3 \le -6$$
$$x_1 + 2x_2 + 5x_3 \le 4$$
$$-x_1 - 2x_2 - 5x_3 \le -4,$$
$$x_1, x_2, x_3 \ge 0.$$

The matrix form of the above problem is

$$\text{Max. } Z = (2, 3, 1)(x_1, x_2, x_3)$$
subject to

$$\begin{bmatrix} 4 & 3 & 1 \\ -4 & -3 & -1 \\ 1 & 2 & 5 \\ -1 & -2 & -5 \end{bmatrix} \begin{bmatrix} x_1 \\ x_2 \\ x_3 \end{bmatrix} \le \begin{bmatrix} 6 \\ -6 \\ 4 \\ -4 \end{bmatrix},$$

Now the dual of the given primal is

$$\text{Min. } Z_D = (6, -6, 4, -4)(y_1', y_1'', y_2', y_2'')$$
$$= 6(y_1' - y_1'') + 4(y_2' - y_2'')$$
subject to

$$\begin{bmatrix} 4 & -4 & 1 & -1 \\ 3 & -3 & 2 & -2 \\ 1 & -1 & 5 & -5 \end{bmatrix} \begin{bmatrix} y_1' \\ y_1'' \\ y_2' \\ y_2'' \end{bmatrix} \ge \begin{bmatrix} 2 \\ 3 \\ 1 \end{bmatrix}.$$

or $4(y_1' - y_1'') + y_2' - y_2'' \geq 2$

$3(y_1' - y_1'') + 2(y_2' - y_2'') \geq 3$

$(y_1' - y_1'') + 5(y_2' - y_2'') \geq 1,$

$y_1', y_1'', y_2', y_2'' \geq 0.$

Hence the required dual is

Max. $Z_D = -5y_1 - 3y_2 + 4y_3$

subject to $-y_1 - y_2 \leq 1$

$3y_1 + 2y_2 + 2y_3 \leq 1$

$-4y_1 - y_3 = 1,$

$y_2, y_3 \geq 0$ y_1 is unrestricted.

Substituting $y_1 = y_1' - y_1''$, $y_2 = y_2 - y_2''$, the required dual is

Min $Z_D = 6y_1 + 4y_2$

subject to $4y_1 + y_2 \geq 2$

$3y_1 + 2y_2 \geq 3$

$y_1 + 5y_2 \geq 1.$

y_1, y_2 are unrestricted in sign.

Example 6:

Write the dual of the following problem

Max. $Z = 3x_1 + 5x_2 + 7x_3$

subject to $x_1 + x_2 + 3x_3 \leq 10$

$4x_1 - x_2 + 2x_3 \geq 15,$

$x_1, x_2 \geq 0$, x_3 *is unrestricted.*

Solution:

First we shall write the given L.P.P. in the standard primal form.

Since the given problem is of maximization therefore all the constraints should have the sign $\leq$.

Substituting $x_3 = x_3' - x_3''$, the standard primal form of the problem is

Max. $Z = 3x_1 + 5x_2 + 7(x_3' - x_3'')$

subject to

$$x_1 + x_2 + 3x_3' - 3x_3'' \le 10$$

$$-4x_1 + x_2 - 2x_3 + 2x_3'' \le -15$$

$$x_1, x_2, x_3', x_3'', \ge 0.$$

The matrix form of the above problem is

$$\text{Max. } Z = (3, 5, 7, -7)\,(x_1, x_2, x_3', x_3'')$$

subject to

$$\begin{bmatrix} 1 & 1 & 3 & -3 \\ -4 & 1 & -2 & 2 \end{bmatrix} \begin{bmatrix} x_1 \\ x_2 \\ x_3' \\ x_3'' \end{bmatrix} \le \begin{bmatrix} 10 \\ -15 \end{bmatrix}.$$

Now the dual of the given problem is

$$\text{Min. } Z_D = (10, -15)\,(y_1, y_2)$$

subject to

$$\begin{bmatrix} 1 & -4 \\ 1 & 1 \\ 3 & -2 \\ -3 & 2 \end{bmatrix} \begin{bmatrix} y_1 \\ y_2 \end{bmatrix} \ge \begin{bmatrix} 3 \\ 5 \\ 7 \\ -7 \end{bmatrix}$$

or $\quad \text{Min. } Z_D = 10y_1 - 15y_2$

subject to

$$y_1 - 4y_2 \ge 3$$

$$y_1 + y_2 \ge 5$$

$$3y_1 - 2y_2 \ge 7$$

$$-3y_1 + 2y_2 \ge -7 \text{ or } 3y_1 - 2y_2 \le 7_1$$

$$y_1, y_2 \ge 0.$$

Thus the required dual is

$$\text{Min. } Z_D = 10y_1 - 15y_2$$

$$y_1 - 4y_2 \ge 3$$

$$y_1 + y_2 \ge 5$$

$$3y_1 - 2y_2 = 7$$

$$y_1, y_2 \ge 0.$$

EXERCISE

Find the dual of the following linear programming problems:

1. Max. $Z = x_1 - x_2 + 3x_3$

   subject to $x_1 + x_2 + x_3 \leq 10$

   $2x_1 - x_3 \leq 2$

   $2x_1 - 2x_2 + 3x_3 \leq 6$

   $x_1, x_2, x_3 \geq 0.$

2. Max. $Z = 3x_1 + 5x_2 + 4x_3$

   subject to $2x_1 + 3x_2 \leq 8$

   $2x_2 + 5x_3 \leq 10$

   $3x_1 + 2x_2 + 4x_3 \leq 15$

   $x_1, x_2, x_3 \geq 0.$

3. Max. $Z = x_1 + 3x_2$

   subject to $3x_1 + 2x_2 \leq 6$

   $3x_1 + x_2 = 4,$

   $x_1, x_2 \geq 0.$

4. Min. $Z = 3x_1 + x_2$

   subject to $2x_1 + 3x_2 \geq 2$

   $x_1 + x_2 \geq 1,$

   $x_1, x_2 \geq 0.$

5. Min. $Z = 4x_1 + 6x_2 + 18x_3$

   subject to $x_1 + 3x_3 \geq 3$

   $x_2 + 2x_3 \geq 5,$

   $x_1, x_2, x_3 \geq 0.$

6. Max. $Z = 3x_1 + 4x_2$

   subject to $2x_1 + 6x_2 \leq 16$

   $5x_1 + 2x_2 \geq 20,$

   $x_1, x_2 \geq 0.$

7. Min. $Z = 2x_1 + 3x_2 + 4x_3$

   subject to

   $2x_1 + 3x_2 + 5x_3 \geq 2$

   $3x_1 + x_2 + 7x_3 = 3$

$x_1 + 4x_2 + 6x_3 \leq 5$

$x_1, x_2 \geq 0$, x_3 is unrestricted.

8. Min. $Z = 2x_1 + 2x_2 + 4x_3$

subject to

$2x_1 + 3x_2 = 5x_3 \geq 2$

$3x_1 + x_2 + 7x_3 \leq 3$

$x_1 + 4x_2 + 6x_3 \leq 5,$

$x_1, x_2, x_3 \geq 0.$

9. Min. $Z = x_1 - 3x_2 - 2x_3$

subject to

$3x_1 - x_2 + 2x_3 \leq 7$

$2x_1 - 4x_2 \geq 12$

$-4x_1 + 3x_2 + 8x_3 = 10$

$x_1, x_2 \geq 0$, x_3 is unrestricted.

10. Min. $Z = x_1 + x_2 + x_3$

subject to $x_1 - 3x_2 + 4x_3 = 5$

$x_1 - 2x_2 \leq 3$

$2x_2 - x_3 \geq 4.$

$x_1, x_2 \geq 0$, x_2 is unrestricted.

11. Min. $Z = 7x_1 + 3x_2 + 8x_3$

subject to $8x_1 + 2x_2 + x_3 \geq 3$

$3x_1 + 6x_2 + 4x_3 \geq 4$

$4x_1 + x_2 + 5x_3 \geq 1$

$x_1 + 5x_2 + 2x_3 \geq 7,$

$x_1, x_2, x_3 \geq 0.$

12. Max. $Z = 3x_1 + x_2 + x_3 - x_4$

subject to $x_1 + 5x_2 + 3x_3 + 4x_4 \leq 5$

$x_1 + x_3 = -1$

$x_3 - x_4 \geq -5,$

$x_1, x_2, x_3, x_4 \geq 0.$

13. Max. $Z = 3x_1 + x_2 + 4x_3 + x_4 + 9x_5$

subject to $4x_1 - 5x_2 - 9x_3 + x_4 - 2x_5 \leq 6$

$2x_1 + 3x_2 + 4x_3 - 5x_4 + x_5 \le 9$

$x_1 + x_2 - 5x_3 - 7x_4 + 11x_5 \le 10,$

$x_1, x_2, x_3, x_4, x_5 \ge 0.$

14. Max. $Z = 3x_1 + x_2 + 2x_3 - x_4$

subject to $2x_1 - x_2 + 3x_3 + x_4 = 1$

$x_1 + x_2 - x_3 + x_4 = 3,$

$x_1, x_2, x_3 \ge 0$, x_4 is unrestricted in sign.

ANSWERS

1. Min. $Z_D = 10y_1 + 2y_2 + 6y_3$

subject to $y_1 + 2y_2 + 2y_3 \ge 1$

$y_1 - 2y_3 \ge -1$

$y_1 - y_2 + 3y_3 \ge 3,$

$y_1, y_2, y_3 \ge 0.$

2. Min. $Z_D = 8y_1 + 10y_2 + 15y_3$

subject to $2y_1 + 3y_3 \ge 3$

$3y^1 + 2y_2 + 2y_3 \ge 5$

$5y_2 + 4y_3 \ge 4,$

$y_1, y_2, y_3 \ge 0.$

3. Min. $Z_D = 6y_1 + 4y_2$

subject to $3y_1 + 3y_2 \ge 1$

$2y_1 + y_2 \ge 3,$

$y_1 \ge 0$, y_2 is unrestricted.

4. Max. $Z_D = 2y_1 + y_2$

subject to $2y_1 + y_2 \le 3$

$3y_1 + y_2 \le 1,$

$y_1, y_2 \ge 0.$

5. Max. $Z_D = 3y_1 + 5y_2$

subject to $y_1 \le 4$

$y_2 \le 6$

$3y_1 + 2y_2 \le 18,$

$y_1, y_2, y_3 \geq 0.$

6. Min. $Z_D = 16y_1 - 20y_2$

subject to $2y_1 - 5y_2 \geq 3$

$6y_1 - 2y_2 \geq 4,$

$y_1, y_2 \geq 0.$

7. Max. $Z_D = 2y_1 - 3y_2 - 5y_3$

subject to $2y_1 - 3y_2 - y_3 \leq 2.$

$3y_1 - y_2 - 4y_3 \leq 3$

$5y_1 - 7y_2 - 6y_3 = 4,$

$y_1, y_3 \geq 0$, y_2 is unrestricted.

8. Max. $Z_D = 2y_1 - 3y_2 - 5y_3$

subject to $2y_1 - 3y_2 - y_3 \leq 2$

$3y_1 - y_2 - 4y_3 \leq 2$

$5y_1 - 7y_2 - 6y_3 \leq 4,$

$y_1, y_2, y_3 \geq 0.$

9. Min. $Z_D = -7y_1 + 12y_2 + 10y_3$

subject to $-3y_1 + 2y_2 - 4y_3 \leq 1$

$-y_1 + 4y_2 - 3y_3 \geq 3$

$2y_1 - 8y_3 = 2,$

$y_1, y_2 \geq 0$, y_3 is unrestricted.

10. Max. $Z_D = 5y_1 + 3y_2 + 4y_3$

subject to $y_1 - y_2 \leq 1$

$-3y_1 + 2y_2 + 2y_3 \leq 1$

$4y_1 - y_3 = 1,$

$y_2, y_3 \geq 0$, y_1 is unrestricted.

11. Max. $Z_D = 3y_1 + 4y_2 + y_3 + 7y_4$

subject to $8y_1 + 3y_2 + 4y_3 + y_4 \leq 7$

$2y_1 + 6y_2 + y_3 + 5y_4 \leq 3$

$y_1 + 4y_2 + 5y_3 + 2y_4 \leq 8$

$y_1, y_2, y_3, y_4 \geq 0.$

12. Min. $Z_D = 5y_1 - y_2 + 5y_3$

subject to $y_1 + y_2 \geq 3$

$5y_1 + y_2 \geq 1$

$3y_1 - y_3 \geq 1$

$4y_1 + y_3 \geq -1,$

$y_1, y_3 \geq 0$, y_2 is unrestricted.

13. Min. $Z_D = 6y_1 + 9y_2 + 10y_3$

subject to $4y_1 + 2y_2 + y_3 \geq 3$

$-5y_1 + 3y_2 + y_3 \geq 1$

$-9y_1 + 4y_2 - 5y_3 \geq 4$

$y_1 - 5y_2 - 7y_3 \geq 1$

$-2y_1 + y_2 + 11y_3 \geq 9$

$y_1, y_2, y_3 \geq 0.$

14. Min. $Z_D = y_1 + 3y_2$

subject to $2y_1 + y_2 \geq 3$

$-y_1 + y2 \geq 1$

$3y_1 - y_2 = 2$

$y_1 + y_2 = -1,$

y_1, y_2 are unrestricted.

UNBALANCED ASSIGNMENT PROBLEM

In case the number of tasks (jobs) is not equal to the number of facilities (persons), the assignment problem is called an *unbalanced assignment problem.* Thus the cost matrix of an unbalanced assignment problem is not a square matrix. For the solution of such problem we add the dummy (fictitious) rows or columns to the given matrix with zero costs to form it a square matrix. Then the usual assignment algorithm can be applied to this resulting baseness assignment problem.

Example 1:

A company is faced with the problem of assigning six different machines to five different jobs. The costs are estimated as follows (in hundreds of rupees):

		Jobs				
		1	*2*	*3*	*4*	*5*
	1	*2.5*	*5*	*1*	*6*	*1*
	2	*2*	*5*	*1.5*	*7*	*3*
Machines	*3*	*3*	*6.5*	*2*	*8*	*3*
	4	*3.5*	*7*	*2*	*9*	*4.5*
	5	*4*	*7*	*3*	*9*	*6*
	6	*6*	*9*	*5*	*10*	*6*

Solve the problem assuming that the objective is to minimize the total cost.

Solution:

Since the matrix is not square, it is an unbalanced assignment problem. We introduce one fictitious job (6th column with zero costs) to get a square matrix. Further the matrix involves elements is decimal. We can make then complete numbers 'by multiplying each element of the cost matrix by 2. The new modified cost matrix is shown in the following table.

	1	2	3	4	5	6
1	5	10	2	12	2	0
2	4	10	3	14	6	0
3	6	13	4	16	6	0
4	7	14	4	18	9	0
5	8	14	6	18	12	0
6	12	18	10	20	12	0

Applying the usual procedure, finally we get the following table:

	1	2	3	4	5	6
1	2	0	0	0	0	4
2	0	0	1	2	2	4
3	0	1	0	2	0	2
4	1	2	0	4	3	2
5	0	0	0	2	4	0
6	4	4	4	4	4	0

Giving the zero assignments, we get the following tables:

	1	2	3	4	5	6
1	2	⊗	⊗	(0)	⊗	4
2	(0)	⊗	1	2	2	4
3	⊗	1	⊗	2	(0)	2
4	1	2	(0)	4	3	2
5	⊗	(0)	⊗	2	4	⊗
6	4	4	4	4	4	(0)

	1	2	3	4	5	6
1	2	⊗	⊗	(0)	⊗	4
2	⊗	(0)	1	2	2	4
3	⊗	1	⊗	2	(0)	2
4	1	2	(0)	4	3	2
5	(0)	⊗	⊗	2	4	⊗
6	4	4	4	4	4	(0)

Thus the original problem has two alternative optimum solution:

(i) $1 \rightarrow 4$, $2 \rightarrow 1$, $3 \rightarrow 5$, $4 \rightarrow 3$, $5 \rightarrow 2$

(ii) $1 \rightarrow 4$, $2 \rightarrow 2$, $3 \rightarrow 5$, $4 \rightarrow 3$, $5 \rightarrow 1$.

The 6th machine will be left unassigned.

From the original matrix, the total minimum cost = 20 *i.e.*, Rs. 2000 (in each case).

Example 2:

A department head has four tasks to be performed and three subordinates. The subordinates differ in efficiency. The estimates of the time, each subordinate would take to perform, are given below in the matrix. How should he allocate the tasks, one to each man, so as to minimize the total man hours?

		Subordinates		
		1	*2*	*3*
Tasks	*I*	*9*	*26*	*15*
	II	*13*	*27*	*6*
	III	*35*	*20*	*15*
	IV	*18*	*30*	*20*

Solution:

Since the matrix is not square, it is an unbalanced assignment problem. We introduce one fictitious subordinate (4th column with zero costs) to get a square matrix. Thus the resulting matrix is shown in the following table. Now the problem can be solved by usual method.

	1	2	3	4
I	9	26	15	0
II	13	27	6	0
III	35	20	15	0
IV	18	30	20	0

Step 1. Subtracting the minimum element of each row from every element of the corresponding row and then subtracting the minimum element of each column from every element of the corresponding column, the matrix reduces to

	1	2	3	4
I	0	6	9	0
II	4	7	0	0
III	26	0	9	0
IV	9	10	14	0

Step 2. Make the 'zero assignments' in usual manner. The illustration is shown in the following table.

	1	2	3	4
I	(0)	6	9	⊗
II	4	7	(0)	⊗
III	26	(0)	9	⊗
IV	9	10	14	(0)

Since in the table every row and every column have one assignment, so we have the complete optimal zero assignment as

I → 1, II → 3, III → 2.

Take IV will be left unassigned.

From the original matrix, the total time (man hours) = 9 + 6 + 20 = 35.

Example 3:

In a machine shop a supervisor wishes to assign five jobs among six machines any one of the job can be processed completely by any one of the machines the cost (in Rs.) of processing any jobs on any machine is given below.

Job

	A	*B*	*C*	*D*	*E*	*F*
1	*13*	*13*	*16*	*23*	*19*	*9*
2	*11*	*19*	*26*	*16*	*17*	*18*
3	*12*	*11*	*4*	*9*	*6*	*10*
4	*7*	*15*	*9*	*14*	*14*	*13*
5	*9*	*13*	*12*	*8*	*14*	*11*

The assignment of jobs to machines must be one-to-one basis. Assign the jobs to machines so that the total cost is minimum. Find the minimum total cost.

Solution:

We add a dummy sixth row (job 6) in the cost matrix so as to get the following balanced assignment problem:

	A	B	C	D	E	F
1	13	13	16	23	19	9
2	11	19	26	16	17	18
3	12	11	4	9	6	10
4	7	15	9	14	14	13
5	9	13	12	8	14	11
6	0	0	0	0	0	0

Now following the usual procedure of solving an assignment problem, an optimum assignment is obtained as follows:

$1 \to F,\ 2 \to A,\ 3 \to E,\ 4 \to C,\ 5 \to D$

with total minimum cost as

Rs. (9 + 11 + 6 + 9 + 8), *i.e.*, Rs. 43.

A company has 4 machines on which to do 3 jobs. Each job can be assigned to one and only one machine. The cost of each job on each machine is given in the following table:

Machine

Job	W	X	Y	Z
A	18	24	28	32
B	8	13	17	19
C	10	15	19	22

What are the job assignments which will minimize the cost?

A department head has six jobs and five subordinates. The subordinates differ in their efficiency and the tasks differ. In third intrinsic difficulty. The department head estimates the time each man would take to perform each task as given in the effectiveness matrix below:

Man	Task A	B	C	D	E	F
1	20	15	26	40	32	12
2	15	32	46	26	28	20
3	11	15	2	12	6	14
4	8	24	12	22	22	20
5	12	20	18	10	22	15

Only one task can be assigned to one man determine how should the job be allocated so as to minimize the total man hour. Find the minimum total man hours.

THE MAXIMAL ASSIGNMENT PROBLEM

We have discussed a problem of minimizing the total cost. Sometimes the assignment problem deals with the maximization of an objective function rather than to minimize it. For example, the problem may be to assign persons to jobs in such a way that the expected profit is maximum. This problem may be solved easily by first converting it to a minimization problem and then we can use the usual procedure of assignment algorithm. This conversion can be very easily done by modifying the given profit matrix to the cost matrix in either of the two ways.

(i) *Select the greatest element of the given profit matrix and then subtract each element of the matrix from this greatest element to get the modified matrix.*

For, if $[c_{ij}]$ is the given profit matrix and c_{rk} is the greatest element of this matrix then the modified matrix will be $[c_{ij}']$, where $c_{ij}' = c_{rk} - c_{ij}$. It can be shown that if $x_{ij} = x_{ij}$ maximizes $Z = \Sigma\Sigma c_{ij}\, x_{ij}$, then $x_{ij} = X_{ij}$ minimizes the function $Z' = \Sigma\Sigma c_{ij}'\, x_{ij}$, where $c_{ij}' = c_{rk} - c_{ij}$. It follows from the relation

$$Z' = \Sigma\Sigma c_{ij}'\, x_{ij} = \Sigma\Sigma\,(c_{rk} - c_{ij})\, x_{ij} = \Sigma\Sigma c_{rk}\, x_{ij} - \Sigma\Sigma c_{ij}\, x_{ij} = nc_{rk} - Z.$$

(ii) *Place minus sign before each element of the profit matrix to get the modified matrix.*

In this case if $[c_{ij}]$ is the given profit matrix then the modified matrix will be $[c_{ij}']$ where $c_{ij}' = -c_{ij}$. It can be shown that if $x_{ij} = X_{ij}$ maximizes $Z = \Sigma\Sigma c_{ij}\, x_{ij}$ then $x_{ij} = X_{ij}$ minimizes $Z' = \Sigma\Sigma c_{ij}'\, x_{ij}$.

Example 1:

A company has four territories open and four salesmen available for assignment. The territories are not equally rich in their sales potential; is estimated that a typical salesman operating in each territory would bring in the following annual sales:

Territory	:	I	II	III	IV
Annual sales (Rs)	:	60,000	50,000	40,000	30,000

Four salesmen are also considered to differ in their ability : it is estimated that, working under the same conditions, their yearly sales would be proportionately as follows:

Salesman	:	A	B	C	D
Proportion	:	7	5	5	4

If the criterion is maximum expected total sales, then intuitive answer is to assign the best salesman to the richest territory, the next best salesman to the second richest, and so on. Verify this answer by the assignment technique.

Solution:

To construct the effectiveness matrix.

The sum of proportions of sales of four salesmen

$$= 7 + 5 + 5 + 4 = 21.$$

If we consider Rs 10,000 as one unit, then A's annual sales in the four territories are

$$\frac{7}{21}\times 6, \frac{7}{21}\times 5, \frac{7}{21}\times 4, \frac{7}{21}\times 3.$$

Similarly we can calculate the annual sales of other salesmen in different territories. In order to avoid fractional values we consider the sales in 21 years.

Thus, the maximum sale matrix is obtained as follows:

Sales in Rs 10,000 →		6	5	4	3
Sales proportion ↓		I	II	III	IV
7	A	42	35	28	21
5	B	30	25	20	15
5	C	30	25	20	15
4	D	24	20	16	12

This is a 'maximization' problem. To convert it into a 'minimization' one let us multiply each element of the above matrix by – 1. Thus resulting matrix becomes:

	I	II	III	IV
A	–42	–35	–28	–21
B	–30	–25	–20	–15
C	–30	–25	–20	–15
D	–24	–20	–16	–12

Now we shall solve the minimization problem by usual assignment algorithm.

Step 1. Subtracting the smallest element of each row from every element of that row and then subtracting the smallest element of each column from every element of that column the reduced matrix is shown in the following table.

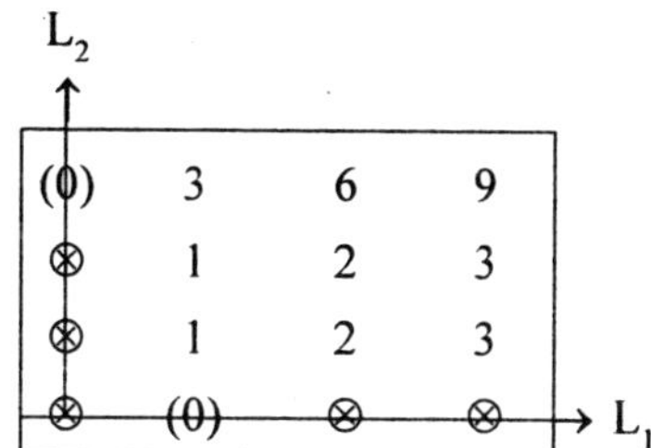

Step 2. Make zero assignments. Since total no. of marked '□' zeros is less than 4, so an optimal assignment is not possible at this stage (as shown in the table of step 1).

Step 3. We draw minimum no. of lines to cover all the zeros.

The minimum element among all uncovered elements is 1.

Then subtract this value 1 from each uncovered element, add 1 to the elements which lie at the intersection of two lines L_1, L_2 and leave other elements unchanged. Thus, the revised matrix is shown in the table.

	L_2	L_3			
	(0)	2	5	8	√ (4)
	⊗	(0)	1	2	√ (5)
	⊗	⊗	1	2	√ (1)
L1	1	⊗	(0)	⊗	
	√ (2)	√ (3)			

Step 4. Giving the 'zero assignments' in the table of step 3, we see that there is no assignment in row 3 and column 4. So we again proceed to draw minimum number of lines to cover all zeros at least once.

Step 5. We find that three lines L_1, L_2, L_3 cover all the zeros. Again minimum element among all uncovered element is 1. Subtracting 1 from all the uncovered elements, adding 1 to the elements which lie at the intersection of the lines L_1, L_2, L_3 and leaving other elements unchanged, we get the following matrix:

0	2	4	7
0	0	0	1
0	0	0	1
2	1	0	0

Step 6. Giving 'zero assignments', we get two assignments as shown in the following tables:

	I	II	III	IV
A	(0)	2	4	7
B	⊗	(0)	⊗	1
C	⊗	⊗	(0)	1
D	2	1	⊗	(0)

	I	II	III	IV
A	(0)	2	4	7
B	⊗	⊗	(0)	1
C	⊗	(0)	⊗	1
D	2	1	⊗	(0)

Thus two optimal solution are

(i) A → I, B → II, C → III, D → IV

(i) A → I, B → III, C → II, D → IV.

From both the solution, it is obvious that the best salesman A is assigned to the richest territory I, the worst salesman D to the poorest territory IV. Salesman B and C being equally good, so they may be assigned to either II or III. This verifies the given intuitive answer.

Example 2:

A company has 5 jobs to be done. The following matrix shows the return in rupees on assigning ith (i = 1, 2, 3, 4, 5) machine to the jth job (j = A, B, C, D, E). Assign the five jobs to the five machines so as to maximize the total expected profit.

		Jobs				
		A	*B*	*C*	*D*	*E*
	1	*5*	*11*	*10*	*12*	*4*
	2	*2*	*4*	*6*	*3*	*5*
Machines	*3*	*3*	*12*	*5*	*14*	*6*
	4	*6*	*14*	*4*	*11*	*7*
	5	*7*	*9*	*8*	*12*	*5*

Solution:

First we shall convect the problem from maximization to minimization. The greatest element of the given matrix is 14. Subtracting all the elements of the given matrix from 14, the modified matrix is obtained. Now we shall follow the usual procedure of solving an assignment problem.

9	3	4	2	10
12	10	8	11	9
11	2	9	0	8
8	0	10	3	7
7	5	6	2	9

Step 1. Subtracting the minimum element of each row from every element of the corresponding row and then subtracting the minimum element of each column from every element of the corresponding column, the modified matrix reduces to

3	1	2	0	7
0	2	0	3	0
7	2	9	0	7
4	0	10	3	6
1	3	4	0	6

Step 2. Now make 'zero assignments'. The following tables show the necessary steps for reaching the optimal assignment : 1 → C, 2 → E, 3 → D, 4 → B, 5 → A with maximum profit of Rs. 50.

3	1	2	(0)	7
(0)	2	⊗	3	⊗
7	2	9	⊗	7
4	(0)	10	3	6
1	3	4	⊗	6

2	⊗	1	⊗	6
⊗	2	(0)	4	⊗
6	1	8	(0)	6
4	(0)	10	4	6
(0)	2	3	⊗	5

1	⊗	(0)	⊗	5
⊗	3	⊗	5	(0)
5	1	7	(0)	5
3	(0)	9	4	5
(0)	3	3	1	5

RESTRICTIONS ON ASSIGNMENT

Sometimes due to some restrictions (technical, legal or others) the assignment of a particular facility to a particular job is not permitted. To overcome such difficulty we assign a very high cost (say, infinite cost) to the corresponding cell, so that the activity will be automatically excluded from the optimal solution.

Example 1:

Four engineers are available to design four projects. Engineer 2 is not competent to design the project B. Given the following time estimates needed to each engineer to design a given project, find how should the engineers be assigned to projects so as to minimize the total design time of four projects.

		Projects			
		A	B	C	D
Engineers	1	12	10	10	8
	2	14	*Not suitable*	15	11
	3	6	10	16	4
	4	8	10	9	7

Solution:

To avoid the assignment 2 → B, we take its time to be very large (say ∞). Then the cost matrix of the resulting assignment problem is shown in the following table:

12	10	10	8
14	∞	15	11
6	10	16	4
8	10	9	7

Now we apply the assignment technique in the usual manner. The following tables show the necessary steps for reaching the optimal solution:

4	2	2	0
3	∞	4	0
2	6	12	(0)
1	3	2	0

3	(0)	⊗	⊗
2	∞	2	(0)
1	4	10	⊗
(0)	1	⊗	⊗

3	(0)	⊗	1
1	∞	1	(0)
(0)	3	9	⊗
⊗	1	(0)	1

Thus the optimal assignment is : 1 → B, 2 → D, 3 → A, 4 → C.

Total minimum cost (time) = 10 + 11 + 6 + 9 = 36.

Example 2:

A job shop has purchased 5 new machines of different types. Theorem 5 available locations in the shop where a machine could be installed. Some of these locations are more desirable than others for particular machines because of their proximity to work centres which would have a heavy work flow to and from these machines. Therefore, the objective is to assign the new machines to the available locations in order to minimize the total cost of material handling. The estimated cost per unit time of materials handling involving each of the machines is given below for the respective locations. Locations 1, 2, 3, 4 and 5 are not considered suitable for machines A, B, C, D and E respectively. Find the optimal solution:

Location (cost in Rs.)

Machine	*1*	*2*	*3*	*4*	*5*
A	×	*10*	*25*	*25*	*10*
B	*1*	×	*10*	*15*	*2*
C	*8*	*9*	×	*20*	*10*
D	*14*	*10*	*24*	×	*15*
E	*10*	*8*	*25*	*27*	×

How will the optimal solution get modified if location 5 is also unsuitable for machine A?

Solution:

Since locations 1, 2, 3, 4 and 5 are not suitable for machines A, B, C, D and E respectively, an extremely large cost (say ∞) should be attached to these locations. The resulting assignment problem cost matrix is shown below:

∞	10	25	25	10
1	∞	10	15	2
8	9	∞	20	10
14	10	24	∞	15
10	8	25	27	∞

∞	2	6	3	0
⊗	∞	0	2	1
⊗	3	∞	0	2
2	0	3	∞	3
⊗	0	6	5	∞

Now following the usual procedure of solving an assignment problem, an optimum assignment is obtained as displayed above. This gives us the optimum assignment as

A → 5, B → 3, C → 4, D → 2, and E → 1

with total minimum cost as Rs. (10 + 10 + 20 + 10 + 10), *i.e.*, Rs. 60.

Now if location 5 is also unsuitable for machine A, we attach an extremely large cost (= ∞) to cell (1, 5). Now applying the assignment algorithm to this modified problem, the following assignment algorithm can easily be obtained:

A → 4, B → 3, C → 5, D → 2, and E → 1

or A → 2, B → 3, C → 4, D → 5, and E → 1

with minimum total cost of Rs. 65.

EXERCISE

1. What is an Assignment problem? Give two areas of its applications.
2. Give the mathematical formulation of an assignment problem.
3. Give an algorithm to solve an assignment problem.
4. Describe the Hungarian method of solving the assignment problem.
5. If (x_{ij}), i = 1, 2, ... ,n; j = 1, 2, ... , n is an optimal solution for an assignment problem with cost (c_{ij}), then it is also optimal for the problem with cost (c_{ij}') when

   $c_{ij}' = c_{ij}$ for i, j = 1, 2, ... , n; j ≠ k

   $c_{ik}' = c_{ik} - A$, where A is a constant.
6. A computer centre has got three expert programmers. The center needs three application programmed to be developed. The head of the computer centre, after studying carefully the programmed to be developed. Estimates the computer time in minutes required by the experts to the application programmes as follows:

		Programme A	Programme B	Programme C
Programmer	1	120	100	80
	2	70	90	110
	3	110	140	120

Assign the programmers to the programmes in such a way that total computer time is least.

7. Find the optimal solution for the assignment problem with the following cost matrix:

	I	II	III	IV	V
A	11	17	8	16	20
B	9	7	12	6	15
C	13	16	15	12	16
D	21	24	17	28	26
E	14	10	12	11	15

8. Find the optimal assignment for the problems with given cost matrix

(i)

	I	II	III	IV
A	5	3	1	8
B	7	9	2	6
C	6	4	5	7
D	5	7	7	6

(ii)

	1	2	3	4
A	10	12	19	11
B	5	10	7	8
C	12	14	13	11
D	8	15	11	9

9. Solve the following minimal assignment problems:

(i)

Person \ Job	1	2	3	4	5
A	8	4	2	6	1
B	0	9	5	5	4
C	3	8	9	2	6
D	4	3	1	0	3
E	9	5	8	9	5

(ii)

Job \ Man	1	2	3	4	5
I	12	8	7	15	4
II	7	9	17	14	10
III	9	6	12	6	7
IV	7	6	14	6	10
V	9	6	12	10	6

(iii)

Machine \ Job	I	II	III	IV	V
A	5	11	10	12	4
B	2	4	6	3	5
C	3	12	5	14	6
D	6	14	4	11	7
E	7	9	8	12	5

(iv)

Job \ Man	I	II	III	IV	V
A	1	3	2	3	6
B	2	4	3	1	5
C	5	6	3	4	6
D	3	1	4	2	2
E	1	5	6	5	4

10. One car is available at each of the stations 1, 2, 3, 4, 5, 6 and one car is required at each of the stations 7, 8, 9, 10, 11, 12. The distances between the various stations are given in the matrix below. How should the cars be despatched so as to minimize the total mileage covered?

	7	8	9	10	11	12
1	41	72	39	52	25	51
2	22	29	49	65	81	50
3	27	39	60	51	32	32
4	45	50	48	52	37	43
5	29	40	39	26	30	33
6	82	40	40	60	51	30

11. A national truck-rental service has a surplus of one truck in each of the cities 1, 2, 3, 4, 5, 6 and a deficit of one truck in each of the cities 7, 8, 9, 10, 11, 12. The distances (in kilometer) between the cities with a surplus and the cities with a deficit are displayed below:

		To					
		7	8	9	10	11	12
	1	31	62	29	42	15	41
	2	12	19	39	55	71	40
	3	17	29	50	41	22	22
From	4	35	40	38	42	27	33
	5	19	30	29	16	20	23
	6	72	30	30	50	41	20

How should the trucks be dispersed so as to minimize the total distance travelled?

12. Five wagons are available at five stations 1, 2, 3, 4, and 5. These are required at five stations I, II, III, IV and V. The mileages between various stations are given by the following table:

13. The owner of a small machine shop has four persons variable to assign to jobs for the day. Five jobs are offered with the expected profit in rupees for each person on each job being as follows:

		Job				
		A	B	C	D	E
	1	6.20	7.80	5.00	10.10	8.20
	2	7.10	8.40	6.10	7.30	5.90
Person	3	8.70	9.20	11.10	7.10	8.10
	4	4.80	6.40	8.70	7.70	8.00

Find the assignment of persons to jobs that will result in a maximum profit. Which job shadow declined?

14. Five operators have to be assigned to five machines. The assignment costs are given in the table below:

		Machine				
		I	II	III	IV	V
	A	5	5	–	2	6
	B	7	4	2	3	4
Operator	C	9	3	5	–	3
	D	7	2	6	7	2
	E	6	5	7	9	1

Operator A cannot operate machine III and operator C cannot operate machine IV. Find the optimal assignment schedule.

15. There are 3 persons P_1, P_2 and P_3 and 5 jobs J_1, J_2, ... J_5. Each person can do only one job and a job is to be done by one person only. Using Hungarian method, find which 2 jobs should be left undone in the following cost minimizing assignment problem.

	J_1	J_2	J_3	J_4	J_5
P_1	7	8	6	5	9
P_2	9	6	7	6	10
P_3	8	7	9	5	6

16. Use the Hungarian method to find which of the two jobs should be left undone when each of the 4 persons will do only one job in the following cost minimizing assignment problem:

		Job					
		J_1	J_2	J_3	J_4	J_5	J_6
	P_1	10	9	11	12	8	5
	P_2	12	10	9	11	9	4
Person	P_3	8	11	10	7	12	4
	P_4	10	7	8	10	10	5

17. The jobs A, B, C are to be assigned to three machines X, Y, Z. The processing costs (Rs.) are as given in the matrix shown below. Find the allocation which will minimize the overall processing cost.

		Machine		
		X	Y	Z
	A	19	28	31
Job	B	11	17	16
	C	12	15	13

18. Use Hungarian method to solve the following cost minimizing assignment problem:

	1	2	3	4
I	20	22	28	15
II	16	20	12	13
III	19	23	14	25
IV	10	16	12	10

ANSWERS

6. 1 → C, 2 → B, 3 → A.

7. A → 1, B → 4, C → 5, D → 3, E → 2; min. cost = 60.

8. (i) A → III, B → IV, C → II, D → 1; min. cost = 16.

(ii) A → 2, B → 3, C → 4, D → 1; min. cost = 38.

9. (i) A → 5, B → 1, C → 4, D → 3, E → 2; min. cost 9.

(ii) I → 3, II → 1, III → 2, IV → 4, V → 5 or I → 3, II → 1, III →4, IV → 2, V → 5.

(iii) A → V, B → IV, C → I, D → II, E → II.

(iv) A → 1, B → IV, C → III, D → II, E → V.

10. 1 → 11, 2 → 8, 3 → 7, 4 → 9, 5 → 10, 6 → 12.

11. 1 → 11, 2 → 8, 3 → 7, 4 → 9, 5 → 10, 6 → 12.

12. 1 → I, 2 → III, 3 → IV, 4 → II, 5 → V; 39 miles.

13. 1 → B, 2 → A, 3 → E, 4 → C, 5 → D.

14. 1 → III, 2 → II, 3 → V, 4 → IV. 5 → I or 1 → IV, 2 → II, 3 → III, 4 → V, 5 → I.

15. A → 2, B → 4, C → 1, D → 3.

16. (1 → 103), (2 → 104), (3 → 101); (4 → 102).
Delhi Delhi Delhi Calcutta

17. 1 → C, 2 → A, 3 → D, 4 → B.

18. A → W, B → X, C → Y or A → W, B → Y, C → X.

19. 2 → A, 3 → B, 4 → D, 6 → C; max. profit = 28.

20. 1 → D, 2 → B, 3 → C, 4 → E; Job A should be declined.

21. A → IV, B → III, C → II, D → I, E → V or A → IV, B → III, C → V, D → II. E → I.

22. $P_1 \rightarrow J_4$, $P_2 \rightarrow J_2$, $P_3 \rightarrow J_5$; Jobs J_1 and J_2 left undone.

23. $P_1 \rightarrow J_5$, $P_2 \rightarrow J_6$, $P_3 \rightarrow J_4$, $P_4 \rightarrow J_2$; Jobs J_1 and J_2 left undone.

24. A → X, B → Y, C → Z.

25. I → 2, II → 4, III → 3, IV → 1 or I → 4, II → 2, III → 3, IV → 1.

SOLVED EXAMPLES

Example 1:

Apply simplex method to solve the following problem

$$Max.\ Z = 30x_1 + 23x_2 + 29x_3$$

subject to $6x_1 = 5x_2 + 3x_3 \leq 26$

$$4x_1 + 2x_2 + 5x_3 \leq 7$$

$$x_1, x_2, x_3 \geq 0.$$

Hence or otherwise find the solution to the dual of the above problem.

Solution:

The given problem is of maximization. Introducing slack variables x_4, x_5 to change the constraint inequalities into equations, we get

$$\text{Max. } Z = 30x_1 + 23x_2 + 29x_3 + 0x_4 + 0x_5,$$

s.t. $6x_1 + 5x_2 + 3x_3 + x_4 = 26$

$4x_1 + 2x_2 + 5x_3 + x_5 = 7,$

$x_1, x_2, \ldots, x_5 \geq 5.$

The starting B.F.S. is

$x_1 = 0, x_2 = 0, x_3 = 0, x_4 = 26, x_5 = 7.$

Now applying simplex method to obtain the optimal solution, we have

		c_j	30	23	29	0	0	Min. ratio
B	c_B	x_B	Y_1	Y_2	Y_3	Y_4	Y_5	x_B/Y_1
Y_4	0	26	6	5	3	1	0	26/6
Y_5	0	7	(4)	2	5	0	1	7/4
$Z = c_B$	$x_B = 0$	Δ_j	30 ↑	23	29	0	0 ↓	x_B/Y_2
Y_4	0	31/2	0	2	–9/2	1	–3/2	31/4
Y_1	30	7/4	1	(1/2)	5/4	0	1/4	7/2
Z=105/2	Δ_j	0 ↓	8 ↑	–17/2	0	–15/2		
Y_4	0	17/2	–4	0	–19/2	1	–5/2	
Y_2	23	7/2	2	1	5/2	0	1/2	
Z=161/2	Δ_j	–16	0	–57/2	0	–23	/2	

In the last simplex table all $\Delta_j \leq 0$ therefore the optimal solution is

$x_1 = 0, x_2 = 7/2, x_3 = 0$ and Max. $Z = 161/2$.

Dual Problem

The given problem is in standard primal form. The dual of the given problem is

$$\text{Min. } Z_D = 26w_1 + 7w_2$$

subject to $6w_1 + 4w_2 \geq 30$

$5w_1 + 2w_2 \geq 23$

$3w_1 + 2w_2 \geq 29,$

$w_1, w_2 \geq 0.$

To read the solution of the dual from the find simplex table of the primal problem.

$w_1 = -\Delta_4 = 0$, $w_2 = -\Delta_5 = 23/2$

and $\quad$ min Z_D max $Z = 161/2$.

Example 2:

Write the dual of the following linear programming problem and hence solve it.

$$Max.\ Z = 3x_1 - 2x_2$$

$$\text{subject to } x_1 \le 4$$

$$x_2 \le 6$$

$$x_1 + x_2 \le 5$$

$$-x_2 \le -1$$

$$x_1, x_2 \ge 0.$$

Solution:

The given problem is in standard primal form. The dual of the given primal is

$$\text{Min. } Z_D = 4w_1 + 6w_2 + 5w_3 - w_4$$

$$\text{subject to } w_1 + w_3 \ge 3$$

$$w_2 + w_3 - w_4 \ge -2,$$

$$w_1, w_2, w_3, w_4 \ge 0.$$

Changing the dual problem to maximization and introducing surplus variable w_5 and slack variable w_6 to change the inequalities into equations, the dual problem becomes

$$\text{Max.. } Z_D' = -4w_1 - 6w_2 - 5w_3 + w_4 + 0w_5 + 0w_6$$

$$\text{subject to } w_1 + w_3 - w_5 = 3$$

$$-w_2 - w_3 + w_4 + w_6 = 2,$$

$$w_1, w_2, \ldots, w_6 \ge 0.$$

Here we have not introduced the artificial variable because in the first constraint equation w_1 will serve the purpose.

The starting B.F.S. is

$$w_1 = 3,\ w_6 = 2.$$

Now applying the simplex method to obtain the optimal solution of the dual problem

B	c_B	c_j / w_B	–4 / W_1	–6 / W_2	–5 / W_3	1 / W_4	0 / W_5	0 / W_6	Min. ratio w_B/W_4
W1	–4	3	1	0	1	0	–1	0	–
W_6	0	2	0	–1	–1	(1)	0	1	2
$Z_D' = -12$		Δ_j	0	–6	–1	1 ↑	–4	0 ↓	
W_1	–4	3	1	0	1	0	–1	0	
W_4	1	2	0	–1	–1	1	0	1	
$Z_D'=-10$		Δ_j	0	–5	0	0	–4	–1	

Since all Δ_j are negative or zero therefore the last table gives the optimal solution of the dual.

The optimal solution of the dual is

$w_1 = 3,\ w_2 = 0,\ w_3 = 0,\ w_4 = 2,$

Min. $Z_D = -$ Max. $Z_D' = 10$.

To read the solution of the primal from the final simplex table of the dual.

The optimal solution of the primal problem is

$x_1 = -\Delta_5 = 4,\ x_2 = -\Delta_6 = 1$ and Max. $Z =$ Min. $Z_D = 10$.

Example 3:

Write the dual of the following problem

Max. $Z = 5x_1 - 2x_2 + 3x_3$

subject to $2x_1 + 2x_2 - x_3 \geq 2$

$3x_1 - 4x_2 \leq 2$

$x_2 + 2x_3 \leq 5,$

$x_1, x_2, x_3 \geq 0.$

Hence or otherwise write the solution of the primal.

Solution:

Writing the given L.P.P. into standard primal form, we get

Max. $Z = 5x_1 - 2x_2 + 3x_3$

subject to

$$-2x_1 - 2x_2 + x_3 \le -2$$
$$3x_1 - 4x_2 \le 3$$
$$x_2 + 2x_3 \le 5,$$
$$x_1, x_2, x_3 \le 0.$$

The dual of the given primal is

$$\text{Min. } Z_D = -2w_1 + 3w_2 + 5w_3$$

subject to

$$-2w_1 + 3w_2 \ge 5$$
$$-2w_1 - 4w_2 + w_3 \ge -2$$
$$w_1 + 2w_3 \ge 3,$$
$$w_1, w_2, w_3 \ge 0.$$

Now to solve the dual problem by simplex method, converting it to maximization problem and multiplying both sides of the second constraint by – 1, we get

$$\text{Max. } Z_D' = 2w_1 - 3w_2 - 5w_3$$

subject to $-2w_1 + 3w_2 + 0w_3 \ge 5$

$$2w_1 + 4w_2 - w_3 \le 2$$
$$w_1 + 0w_2 + 3w_3 \ge 3,$$
$$w_1, w_2, w_3 \ge 0.$$

Introducing surplus variables w_4, w_6, slack variable w_5 and artificial variables A_1, A_2 the above dual problem changes to

$$\text{Max. } Z_D' = 2w_1 - 3w_2 - 5w_3 + 0w_4 + 0w_5 + 0w_6 - MA_1 - MA_2$$

subject to

$$-2w_1 + 3w_2 + 0w_3 - w_4 + A_1 = 5$$
$$2w_1 + 4w_2 - w_3 + 0w_4 + w_5 = 2$$
$$w_1 + 0w_2 + 3w_3 - w_6 + A_2 = 3.$$
$$w_1, w_2, \ldots, w_6 \ge 0.$$

The starting B.F.S. is $w_1 = 0 = w_2 = w_3 = w_4 = w_6$, $w_5 = 2$, $A_1 = 5$, $A_2 = 3$.

Now apply the simplex method to find the optimal solution of dual problem.

		c_j	2	−3	−5	0	0	0	−M	−M	Min. ratio
B	c_B	x_B	W_1	W_2	W_3	W_4	W_5	W_6	A_1	A_2	w_B/W_2
A_1	−M	5	−2	3	0	−1	0	0	1	0	5/3
W_5	0	2	2	(4)	−1	0	1	0	0	0	2/4
A_2	−M	3	1	0	3	0	0	−1	0	1	–
$Z_D' = -8M$		Δ_j	2 − M− 3 + 3M− 5 + 3M				−M	0	−M	0	0 w_B/W_3
A_1	−M	7/2	−7/2	0	3/4	−1	−3/4	0	1	0	14/3
W_2	−3	1/2	1/2	1	−1/4	0	1/4	0	0	0	–
A_2	−M	3	1	0	(3)	0	0	−1	0	1	1
$Z_D' = -\frac{13}{2}M - \frac{3}{2}$		Δ_j	$\frac{7-5M}{2}$	0	$\frac{15M-23}{4}$	−M	$\frac{303M}{4}$	−M	0	0	w_B/W_6
				↑					↓		
A_1	−M	11/4	−15/4	0	0	−1	−3/4	(1/4)	1	11	
W_2	−3	3/4	7/12	1	0	0	1/4	−1/12	0	–	
W_3	−5	1	1/3	0	1	0	0	−1/3	0	–	
$Z_D' = -\frac{11}{4}M - \frac{29}{4}$		Δ_j	$\frac{65-45M}{12}$		0	0	−M	$\frac{3-3M}{4}$	$\frac{3M-23}{12}$ ↓	0 ↓	
W_6	0	11	−15	0	0	−4	−3	1			
W_2	−3	5/3	−2/3	1	0	−1/3	0	0			
W_3	−5	14/3	−14/3	0	1	−4/3	−1	0			
$Z_D' = -\frac{85}{3}$		Δ_j	−70/3	0	0	−23/3	−5	0			

In the last simplex table all $\Delta_j \leq 0$ therefore it gives optimal solution of the dual problem.

The optimal solution of the dual is

$$w_1 = 0, \; w_2 = 5/3, \; w_3 = 14/3$$

and $\quad$ $\text{Min. } Z_D = -\text{ Max. } Z_D' = 85/3.$

Now the optimal solution of the primal is

$$x_1 = -\Delta_4 = 23/3, \; x_2 = -\Delta_5 = 5,$$

$$x_3 = -\Delta_6 = 0$$

and $\quad$ $\text{Max. } Z = \text{Min. } Z_D = 85/3.$

Example 4:

Use duality to solve

Min. $Z = 3x_1 + x_2$

subject to $x_1 + x_2 \geq 1$

$2x_1 + 3x_2 \geq 2,$

$x_1, x_2 \geq 0.$

Solution:

The given L.P.P. is in standard primal form. The dual problem is given by

Max. $Z_D = w_1 + 2w_2$

subject to $w_1 + 2w_2 \leq 3$

$w_1 + 3w_2 \leq 1,\ w_1, w_2 \geq 0.$

Introducing slack variables w_3 and w_4 to change the constraint inequalities into equations, the dual problem becomes

Max. $Z_D = w_1 + 2w_2 + 0w_3 + 0w_4$

subject to $w_1 + 2w_2 + w_3 = 3$

$w_1 + 3w_2 + w_4 = 1,$

$w_1, w_2, w_3, w_4 \geq 0.$

Initial B.F.S. is $w_3 = 3,\ w_4 = 1.$

Now we shall apply simplex method to obtain solution.

		c_j	1	2	0	0	Min ratio
B	c_B	w_B	W_1	W_2	W_3	W_4	w_B/W_2
W_3	0	3	1	2	1	0	3/2
W_4	0	1	1	(3)	0	1	<u>1/3</u>
Z_D=0		Δ_j	1	2 ↑	0	0 ↓	w_B/W_1
W_3	0	7/3	1/3	0	1	–2/3	7
W_2	2	1/3	(1/3)	1	0	1/3	<u>1</u>
Z_D=2/3		Δ_j	1/3 ↑	0 ↓	0	–2/3	
W_3	0	2	0	–1	1	–1	
W_1	1	1	1	3	0	1	
Z_D=1	Δ_j	0	–1	0	–1		

Since all $\Delta_j \le 0$, therefore the dual problem has optimal solution

$$w_1 = 1, w_2 = 0, \text{Max. } Z_D = 1.$$

Now the solution of the original primal problem form the last simplex table of the dual is

$$x_1 = -\Delta_3 = 0, x_2 = -\Delta_4 = 1, \text{min. } Z = \text{max. } Z_D = 1.$$

Example 5:

One unit of product A contributes Rs. 7 and requires 3 units of raw material and 2 hours of labor. One unit of product B contributes Rs. 5 and requires one unit of raw material and one hour of labor. Availability of the raw material at present is 48 units and there are 40 hours of labor.

(i) *Formulate the linear programming problem,*

(ii) *Write the dual and solve it by simplex method. Also find the optimal product mix.*

Solution:

The linear programming problem corresponding to the given information is

$$\text{Max. } Z = 7x_1 + 5x_2$$

subject to $3x_1 + x_2 \le 48$

$$2x_1 + x_2 \le 40$$

$$x_1, x_2 \ge 0.$$

The given L.P.P. is in standard primal form.

The dual problem can be written as

$$\text{Min. } Z_D = 48w_1 + 40w$$

subject to $w_1 + 2w_2 \ge 7$

$$w_1 + w_2 \ge 5,$$

$$w_1, w_2 \ge 0.$$

To solve the dual problem by simplex method, changing the dual objective function to maximization and introducing surplus variables w_3, w_4 and artificial variables w_5, w_6 respectively to change the constraint incredulities into equations. The dual problem can be written as

$$\text{Max. } Z_D' = -48w_1 - 40w_2 + 0w_3 + 0w_4 - Mw_5 - Mw_6$$

subject to $3w_1 + 2w_2 - w_3 + w_5 = 7$

$$w_1 + w_2 - w_4 + w_6 = 5,$$

$w_1, w_2, \ldots, w_6 \geq 0.$

The initial B.F.S. is

$$w_1 = 0 = w_2 = w_3 = w_4, \; w_5 = 7, \; w_6 = 5.$$

Now applying simplex method to obtain the optimal solution of the dual problem.

		c_j	–48	–40	0	0	–M	–M	Min. ratio
B	c_B	w_B	W_1	W_2	W_3	W_4	W_5	W_6	w_B/W_1
W_5	–M	7	(3)	2	–1	0	1	0	7/3
W_6	–M	5	1	1	0	–1	0	1	5
$Z_D' = -12M$		Δ_j	$4M-48$	$3M-40$	–M	–M	0	0	w_B/W_2
			↓				↑		
W_1	–48	7/3	1	(2/3)	–1/3	0	1/3	0	7/2
W_6	–M	8/3	0	1/3	1/3	–1	–1/3	1	8
$Z_D' = -\left(112 + \frac{8}{3}M\right)$		Δ_j	0	$\frac{M}{3} - 8$	–16	–M	$16 - \frac{4M}{3}$	0	
			↓	↑					

		c_j	–48	–40	0	0	–M	–M	Min. ratio
B	c_B	w_B	W_1	W_2	W_3	W_4	W_5	W_6	w_B/W_1
W_2	–40	7/2	3/2	1	–1/2	0	–1/2	0	–
W_6	–M	3/2	–1/2	0	(1/2)	–1	–1/2	1	3
$Z_D' = -\left(140 + \frac{3}{2}M\right)$		Δ_j	$12 - \frac{1}{2}M$	0	$\frac{1}{2}M - 20$	–M	$20 - \frac{3}{2}M$	0	
					↑				↓
W_2	–40	5	1	1	0	–1	0	1	
W_3	0	3	–1	0	1	–2	–1	2	
$Z_D' = -200$		Δ_j	–8	0	0	–40	–M	$40 - M$	

∴ Solution of the dual problem is

$$w_1 = 0, \; w_2 = 5,$$

Min. $Z_D = -$ Max. $Z_D' = 200.$

Hence the solution of the primal problem is

$$x_1 = -\Delta_3 = 0, \; x_2 = -\Delta_4 = 40$$

and Max. $Z = 200$.

Thus the optimal product mix is:

none of the product A and 40 units of product B for a total conscription of Rs. 200.

EXERCISE

1. Find the solution of the following problem by simplex method:

Max. $Z_x = 40x_1 + 50x_2$

subject to $2x_1 + 3x_2 \leq 3$, $8x_1 + 4x_2 \leq 5$, $x_1, x_2 \geq 0$.

Write the dual of the given L.P.P. Also find the solution of the dual problem.

Use principle of duality to solve the following problems:

2. Min. $Z = x_1 - x_2$

subject to $2x_1 + x_2 \geq 2$, $-x_1 - x_2 \geq 1$ and $x_1, x_2 \geq 0$.

3. Max. $Z = 3x_1 + 2x_2$

subject to $2x_1 + x_2 \leq 5$, $x_1 + x_2 \leq 3$ and $x_1, x_2 \geq 0$.

4. Max. $Z - 2x_1 + x_2$

subject to $x_1 + 2x_2 \leq 10$, $x_1 + x_2 \leq 6$

$x_1 - x_2 \leq 2$, $x_1 - 2x_2 \leq 1$ and $x_1, x_2, x_3 \geq 0$.

5. Min. $Z = 2x_1 + 2x_2$

subject to $2x_1 + 4x_2 \geq 1$, $x_1 + 2x_2 \geq 1$.

$2x_1 + x_2 \geq 1$, and $x_1, x_2 \geq 0$.

6. Max. $Z = 3x_1 + 2x_2$

subject to $x_1 + x_2 \geq 1$, $x_1 + x_2 \leq 7$,

$x_1 + 2x_2 \leq 10$, $x_2 \leq 3$ and $x_1, x_2 \geq 0$.

7. Formulate the dual of the given L.P.P. and hence solve it:

Min. $Z = 3x_1 - 2x_2 + 4x_3$

subject to $3x_1 + 5x_2 + 4x_3 \geq 7$, $6x_1 + x_2 + 3x_3 \geq 4$,

$7x_1 - 2x_2 - x_3 \leq 10$, $x_1 - 2x_2 + 5x_3 \geq 3$

$4x_1 + 7x_2 - 2x_3 \geq 2$

and $x_1, x_2, x_3 \geq 0$.

8. Solve the following problem by simplex method:

Max. $Z = 30x_1 + 23x_2 + 29x_3$

subject to $6x_1 + 5x_2 + 3x_3 \le 26$, $4x_1 + 2x_2 + 5x_3 \le 7$,
$x_1, x_2, x_3 \ge 0$.

Also read the solution to the dual of the above problem.

9. Use duality theory to solve the given L.P.P.

Min. $Z = 4x_1 + 3x_2 + 6x_3$

subject to $x_1 + x_3 \ge 2$, $x_2 + x_3 \ge 5$ and $x_1, x_2, x_3 \ge 0$.

10. Solve the following L.P.P. using duality theory

Min. $Z = 50x_1 - 80x_2 - 140x_3$,

subject to $x_1 - x_2 - 3x_3 \ge 4$, $x_1 - 2x_2 - 2x_3 \ge 3$

and $x_1, x_2, x_3 \ge 0$.

11. Apply the principle of duality to solve the following L.P.P.

Max. $Z = 3x_1 - 2x_2$

subject to $x_1 + x_2 \le 5$, $x_1 \le 4$, $1 \le x_2 \le 6$

and $x_1, x_2 \ge 0$.

12. Solve the following primal problem and its dual by simplex method

Max. $Z = 5x_1 + 12x_2 + 4x_3$

subject to $x_1 + 2x_2 + x_3 \le 5$, $2x_1 - x_2 + 3x_3 = 2$, $x_1, x_2, x_3 \ge 0$.

Verify that the solution of the primal can be read from the optimal table of the dual and vice-versa.

13. Use duality theory to solve

Min. $Z = 3x_1 + x_2$

subject to $x_1 + x_2 \ge 1$, $2x_1 + 3x_2 \ge 2$, $x_1, x_2, x_3 \ge 0$.

14. Write the dual of the following primal

Max. $Z = 40x_1 + 35x_2$

subject to $2x_1 + 3x_2 \le 60$, $4x_1 + 3x_2 \le 96$, $x_1, x_2 \ge 0$.

Solve the primal and the dual by simplex method. Compare the optimal solution of the two problems.

15. Use duality to solve the following L.P.P.

Max. $Z = 4x_1 + 3x_2$

subject to $x_1 \le 6$, $x_2 \le 8$, $x_1 + x_2 \le 7$, $3x_1 + x_2 \le 15$, $-x_2 \le 1$ and $x_1, x_2 \ge 0$.

16. Find the dual of the following L.P.P.

Max. $Z = 2x_1 - x_2$

subject to $x_1 + x_2 \le 10, - 2x_1 + x_2 \le 2$

$4x_1 + 3x_2 \ge 12, x_1, x_2, \ge 0.$

Solve the primal problem by simplex method and deduce from it the solution to the dual problem.

17. Find the dual of the following problem

Min. $Z = 6x + 5y + 2x$

subject to $x + 3y + 2z \ge 5, 2x + 2y + z \ge 2,$

$4x - 2y + 3z \ge - 1, x, y, z \ge 0.$

Hence or otherwise solve the primal problem.

18. Use duality to solve the problem

Min. $Z = 10x_1 + 6x_2 + 2x_3$

subject to $- x_1 + x_2 + x_3 \ge 1, 3x_1 + x_2 - x_3 \ge 2$, and $x_1, x_2, x_3 \ge 0.$

19. Write the dual of the following problem

Max. $Z = 2x_1 + 3x_2,$

subject to $2x_1 + 2x_2 \le 10, 2x_1 + x_2 \le 6,$

$x_1 + 2x_2 \le 6, x_1, x_2, \ge 0.$

Solve the primal and then find the solution to the dual.

20. A company makes three products X, Y, Z out of three raw materials A, B, C. The number of units of raw materials required to produce one unit of the product is as given in the following table:

	X	Y	Z
A	1	2	1
B	2	1	4
C	2	5	1

The unit profit contribution of the products X, Y and Z are Rs. 40, 25 and 50 respectively.

The number of units of raw materials available are 36, 60 and 45 respectively.

(i) Determine the product mix that will maximize the total profit.

(ii) Through the final simplex table write the solution to the dual problem.

ANSWERS

1. $x_1 = 3/16$, $x_2 = 7/8$, max. $Z_x = 51.25$

Dual. $w_1 = 15$, $w_2 = 5/4$, min. $Z_D = 51.25$.

2. no feasible solution.

3. $x_1 = 2$, $x_2 = 1$, max. $Z = 8$.

4. $x_1 = 4$, $x_2 = 2$, max. $Z = 10$.

5. $x_1 = 1/3$, $x_2 = 1/3$, min. $Z = 4/3$.

6. $x_1 = 7$, $x_2 = 0$, Max. $Z = 21$.

7. no feasible solution.

8. $x_1 = 0$, $x_2 = 7/2$, $x_3 = 0$, max. $Z = 161/2$

Dual. $w_1 = 0$, $w_2 = 23/2$, min. $Z_w = 161/2$.

9. $x_1 = 0$, $x_2 = 3$, $x_3 = 2$, min. $Z = 21$.

10. no feasible solution.

11. $x_1 = 4$, $x_2 = 1$, max. $Z = 10$.

12. $x_1 = 9/5$, $x_2 = 8/5$, $x_3 = 0$, max. $Z = 28\frac{1}{5}$,

Dual. $w_1 = 29/5$, $w_2 = -2/5$, Min. $Z_D = 28\frac{1}{5}$.

13. $x_1 = 0$, $x_2 = 1$, Min. $Z = 1$.

14. $x_1 = 18$, $x_2 = 8$, max. $Z = 1000$.

Dual. $w_1 = 10/3$, $w_2 = 25/3$, min. $Z_D = 1000$.

15. $x_1 = 4$, $x_2 = 3$, max. $Z = 25$.

16. $x_1 = 10$, $x_2 = 0$, max. $Z = 20$.

Dual. $w_1 = 2$, $w_2 = 0$, $w_3 = 0$, min. $Z_D = 20$.

17. $x_1 = 0$, $x_2 = 0$, $x_3 = 5/2$, min. $Z = 5$.

18. $x_1 = 1/4$, $x_2 = 5/4$, $x_3 = 0$, Min. $Z = 10$.

19. $x_1 = 2$, $x_2 = 2$, max. $Z = 10$.

Dual. $w_1 = 0$, $w_2 = 1/3$, $w_3 = 4/3$, min. $Z_D = 10$.

20. $x = 20$, $y = 0$, $z = 5$, Max. $P = 1050$,

Dual. $r = 0$, $s = 10$, $t = 10$, min. $R = 1050$.

EXERCISE
(Objective Questions)

Fill in the Blanks:

Fill in the blanks " ... ", so that the following statements are complete and correct.

1. If the primal problem is a maximization problem, its dual will be a ... problem.
2. If any constraint in the primal is a perfect equality, the corresponding dual variable is ... is sign.
3. If the primal problem has an unbounded solution, the dual has either no solution or an ... solution.
4. If the primal has a finite optimal solution then the values of the objective functions of the primal and dual are ... .
5. If x is any feasible solution to the primal problem and w is any feasible solution to the dual problem then Z_P ... Z_D where Z_P and Z_D function are the objective of the primal and dual respectively.

Multiple Choice Questions:

Indicate the correct answer for each question by writing the corresponding letter from (a), (b), (c) and (d).

6. In standard primal form if the problem is of maximization, all the constraints involve the sign

   (a) $\geq$ (b) $\leq$

   (c) = (d) unrestricted.

7. If the standard primal is of minimization, all the constraints involve the sign

   (a) $\geq$ (b) $\leq$

   (c) = (d) none of these

8. If a finite optimal feasible solution exists for the primal then the dual has

   (a) unbounded solution

   (b) no solution

   (c) a finite feasible optimal solution

   (d) none of these

9. If the primal problem has an unbounded solution, the dual problem has
   (a) a finite optimal feasible solution
   (b) no solution
   (c) either no solution or an unbounded solution
   (d) none of these

10. If the ith primal variable of the primal is positive, then the ith variable of the dual is
    (a) +ive (b) –ve
    (c) zero (d) unrestricted

11. If both the primal and dual problems have finite optimal solution and Z_P. Z_D are the optimal values of the objective function of the primal and dual respectively then we have
    (a) $Z_P > Z_D$ (b) $Z_P < Z_D$
    (c) $Z_P = Z_D$ (d) none of these.s

True or False:

Write 'T' for true and 'F' for the false statement.

12. The dual of a dual is the primal itself.
13. If the primal problem has a finite feasible solution, the dual problem has no solution.
14. The coefficient matrix of the dual is obtained by transposing the coefficient matrix of the primal.
15. If the primal is a maximization problem, the dual is also a maximization problem
16. Requirement vector of the primal is the price vector of the dual.
17. In a standard primal problem, if all the constraints have the sign ≥, it is a maximization problem.

ANSWERS

1. minimization. **2.** unrestricted. **3.** unbounded.

4. equal. **5.** ≤. **6.** (b). **7.** (a). **8.** (c). **9.** (c).

10. (c). **11.** (c). **12.** T. **13.** F. **14.** T. **15.** F.

16. T. **17.** F.